POLYCRYSTALLINE MATERIALS – THEORETICAL AND PRACTICAL ASPECTS

Edited by **Zachary Todorov Zachariev**

Polycrystalline Materials – Theoretical and Practical Aspects
Edited by Zachary Todorov Zachariev

Published by InTech
Janeza Trdine 9, 51000 Rijeka, Croatia

Publishing Process Manager Adriana Pecar

Technical Editor Teodora Smiljanic

Cover Designer InTech Design Team

Additional hard copies can be obtained from orders@intechopen.com

Polycrystalline Materials – Theoretical and Practical Aspects,
Edited by Zachary Todorov Zachariev
p. cm.
ISBN 978-953-307-934-9

For EU product safety concerns:
IN TECH d.o.o., Prolaz Marije Krucifikse Kozulić 3, 51000 Rijeka, Croatia,
info@intechopen.com or visit our website at intechopen.com.

Meet the editor

Having obtained his degree from Sofia University in 1969, Dr Zachary Todorov Zachariev subsequently went on to become a Doctor of Chemistry (1982) and Doctor of Sciences (2001). From 1989 to 2010 he held the post of Chief of Laboratories at the Institute of General and Inorganic Chemistry during which time he published 51 scientific papers, 3 monographs, 133 citations and secured 12 patens. Dr Zachariev's reputation as a leading specialist in metallurgy has seen him lead and coordinate a number of high profile projects for several international projects, including being the leader on subject 05.06 "Powder Metallurgy" for the Council for Mutual Economic Assistance (CMEA). As a fellow and member of the International Centre for Science & High Technology and ASM International Heat Treating Society, he remains one of the most respected and revered experts within this field, as testified by a wealth of scientific awards throughout is career, including Gold medal "Vermeil", International Exhibition Switzerland'75, Gold medal from International Plovdiv Fair' 87 (Bulgaria), Bronz medal from Exhibition "East-West Europe intellect", Sofia 1998; and the Medal "Kurnakov" from Soviet Academy of Sciences.

Contents

Preface

Polycrystalline structures are conglomerates of a large number of crystals irregularly situated, yet bound to each other strongly enough to behave as a whole. As the size and the shape of these crystals are irregular too, the latter are called grains or crystallites. The boundary surfaces connecting the grains have crystal structures that are not identical to those of the adjacent crystallites. They are distortions, which allow a smooth transition between the structures within the grains in contact. The mosaic of these borderline regions represents an extended block of two dimensional imperfections.

A mechanical loading of these materials leads to deformations. With small loadings, the deformation is elastic, slip and elastic (Young's) modulus being its characteristics. The Young's modulus is an important parameter of the polycrystal materials determining its resistance to deformation.

For higher loadings, the deformation becomes plastic. Theoretical studies, experimental data, as well as practical observations show that this type of deformation involves slippage in the material and active participation of its two-dimensional imperfections. With lower temperatures (less than 0.4 T_m for metals and 0.6 T_m for alloys, T_m denoting the melting point) slippage does not occur uniformly, but remains confined to smaller regions, which appear successively. At higher temperatures, the critical break tension drops down. Thus, smaller loadings prove sufficient to bring about deformation effects, such as dislocations slip, twinning, sliding of grain boundaries, etc.

Stress level, stress rate, and temperature are the parameters characterizing the plastic deformation of polycrystalline materials.

In these materials, the macroscopic value of their parameters is a mean value, resulting from taking the average over domains comprising a considerable number of grains with usually different orientations. In this way, they turn out isotropic as compared to the monocrystals, in which sharp anisotropy is observed. In special cases, the orientations of the grains show more or less preferred directions, leading to anisotropy. To the best of my knowledge, one of them – polycrystalline tungsten, is dealt with for the first time in specialized literature by Dr. P. V. Galptshyan in the chapter: "Strength of Polycrystalline Materials".

The book "Polycrystalline Materials" presents theoretical and practical investigations of some polycrystals materials.

Section I "Plastic Deformation, Strength and Grain-Scale Approaches to Polycrystals" comprises the following three chapters:

"Scale-Bridging Modelling of Plastic Deformation and Damage Initiation in Polycrystals" by Dr. Anxin Ma and Prof. Alexander Hartmaier.

This chapter reviews the current state of modelling the phenomenon of plastic deformation in its various aspects. The authors' analysis involves the macro-, meso-, micro-, and atomic scales using the finite element method, representative volume element approaches, the dislocation dynamics method, and molecular dynamics simulation. Steels have been used to generate realistic microstructures for all multiphase polycrystalline materials studied. Possible approaches in order to bridge the different length scales have been discussed and successful multiscale modelling applications reported. On the basis of recent constitutive models provided by continuum mechanics, the authors have elaborated a number of numerical procedures aiming at the integration of results obtained for several length scales. In this way, they were able to build representative volume element (RVE) models for the mechanical behavior of materials, which are heterogeneous, with respect to the nature of grains and phases they are made of. Once a RVE for a given microstructure is constructed and the critical deformation and damage mechanisms are included into the constitutive relations, this RVE can be applied to calculate stress-strain curves and other mechanical data. The advantage of this approach is that the effects of grain size and strength of the grain boundaries on the macroscopic mechanical response of a material can be predicted.

"Strength of a Polycrystalline Material" by Prof. P. A.Galptshyan

It is shown that due to the greater concentration of stresses in it, the polycrystalline material has a strength less than that of a monocrystal made of the same substance. Hence, in order to enhance its strength, one has to reduce the stresses in the material. A remarkable case is polycrystalline tungsten, whose elastic anisotropy factor proves zero. This kind of tungsten is known to be a most durable substance, more so than the tungsten monocrystals and even the diamond. For example, the ultimate strength under tension of unanealed wires of polycrystalline tungsten is in the range of 1800 MPa to 4150 Mpa, depending on the diameter of the wire. For diamond monocrystals, it equals 1800 MPa at 20^0 C. It is worth noting that a correspondence has been found for polycrystalline metals between their ultimate strength and their modulus of elasticity: the two parameters are changing in the same direction.

"Grain-Scale Modelling Approaches for Polycrystalline Aggregates" by Dr. I. Simonovski and Dr. I. Cizelj

It has been shown that polycrystalline aggregates their microstructure, which plays an important role in the evolution of stresses and strains, and in the development of damage processes, such as small cracks in the microstructure and fatigue. Damage initialization and evolution are directly influenced by the locally anisotropic behavior of the microstructure, as determined by the combination of random grain shapes and sizes, different crystallographic orientations, inclusions, voids, and other microstructural features. For the bulk of a grain, constitutive models assuming pure anisotropic elasticity, as well as anisotropic elasticity in combination with crystal plasticity have been used. Analytical models for grain geometry, in view of calculating the properties of crystalline aggregates, involve 2D and 3D Voronoi tesselations, whereas the method of X-ray diffraction contrast tomography was utilized to measure these properties and make a 3D characterization of the grains.

Several cases of 3D Voronoi tesselations, and two cases of stainless steel wire have been treated. Grain boundaries were explicitly modeled, using the cohesive zone approach, with finite elements of zero physical thickness. Initialization and early development of grain boundary damage, with respect to stainless steel, were traced numerically for several constitutive laws. Differences obtained in the results are small when the approach of anisotropic elasticity is compared to combining the latter with crystal plasticity, except for the computation time required- more than two times longer for the second approach.

Section II "Methods of Synthesis, Structural Properties Characterization, and Applications of Some Polycrystalline Materials" includes the following four chapters:

"Nanocrystalline NASICON Materials - Structure and Electrical Properties" by Dr. Lakshmi Vijayan and Prof. G. Govindaraj.

The chapter deals with an important class of solid electrolytes – sodium (NA) super (S)-ionic (I) conductors (CON): $A_x B_y (PO_4)_3$, where A denotes an alkali metal ion and B denotes a multivalent metal ion. They are widely tested in energy applications, e.g. electric vehicles, and having the advantage of high ionic conductivity together with the stability of the phosphate units. The authors have investigated the correlation between ionic conduction and phase symmetry for a family of NASICONs, comprising $LiTi_2(PO_4)_3$ and $A_3M_2(PO_4)_3$ where A = Li, Na and M = Cr, Fe) . Structural characterization was obtained by spectroscopic and diffraction techniques, while mobile ions characterization proceeded through impedance spectra. Application of the materials studied has been discussed as well.

"Structural Characterization of New Perovskites" by Dr. A. D. Lozano – Gorrin.

The author discusses some general features of the perovskite-type materials, including relatively new methods of their preparation (solution combustion, sonochemical procedures, microwave assisted synthesis) as well as characterization of their structure and physical properties by a variety of diffraction techniques (X-rays-, electron- and neutron diffraction).

"*Controlled Crystallization of Isotactic Polypropylene Based on Alpha/Beta Compounded Nucleating agents - From Theory to Practice*" by Prof. Zhong Xin and Dr.Yaoqi Shi.

Isotactic polypropylene (iPP), one of the most important thermoplastic polymers, exhibits very interesting polymorphic behavior. Its different crystalline forms have different optical and mechanical properties. In this respect, alpha/beta compounded nucleating agents for polypropylene attract more and more attention.

Three kinds of well studied alpha/beta compounded NAs (phosphate/amide, sorbitol/amide, and phosphate/carboxylate) have been reviewed by discussing their influence on the crystallization kinetics, crystallization morphologies, and mechanical properties of iPP. The results show that these three NAs are able to not only increase the crystallization temperature of iPP, but also to shorten its crystallization half-time. Consequently, they are able to considerably reduce the molding cycle time. It has also been found that the type of nucleation of the polymer could be changed, while the geometry of its crystal growth remains the same.

"*Influence of Irradiation on Mechanical Properties of Materials*" by Prof. V. I. Krasilnikov.

This chapter discusses substantial changes in the mechanical properties of materials, radiation embrittlement, and hardening being two of its most common and important effects. Both of them depend on the temperature of the irradiated material.

The author has proposed a model, based on the interaction of vacancies with interstitial barriers in order, to explain and investigate the saturation of the dependence of yield strength on radiation dosage. In the framework of this model, equations describing the evolution of barrier densities, as well as yield strength have been obtained in analytical form. It has been shown that with increasing the intensity of the barrier recombination processes, the yield strength of the irradiated material decreases, the dependence being nonlinear. In the case of radiation hardening, the model proves valid for both low and large doses.

Another model quantitively describing the dependence of the yield strength of irradiated materials on their temperature has also been introduced and applied. The results show its usefulness in dealing with the processes of plastic deformation under irradiation. Some implications about materials used in the construction of nuclear reactors have been discussed.

The research has been carried out to increase the lifetime of III and IV generation reactors and practical ITER-materials.

Prof. D.Sc.Eng. Zachari Zachariev
Institute of Polymers
Bulgarian Academy of Sciences, Sofia
Bulgaria

Part 1

Plastic Deformation, Strength and Grain – Scale Approaches to Polycrystals

1

Scale Bridging Modeling of Plastic Deformation and Damage Initiation in Polycrystals

Anxin Ma and Alexander Hartmaier
Interdisciplinary Centre for Advanced Materials Simulation, Ruhr-University Bochum
Germany

1. Introduction

Plastic deformation of polycrystalline materials includes dislocation slip, twinning, grain boundary sliding and eigenstrain produced by phase transformations and diffusion. These mechanisms are often alternative and competing in different loading conditions described by stress level, strain rate and temperature. Modelling of plasticity in polycrystalline materials has a clear multiscale character, such that plastic deformation has been widely studied on the macro-scale by the finite element methods, on the meso-scale by representative volume element approaches, on the micro-scale by dislocation dynamics methods and on the atomic scale by molecular dynamics simulations. Advancement and further improvement of the reliability of macro-scale constitutive models is expected to originate from developments at microstructural or even smaller length scales by transfering the observed mechanisms to the macro-scale in a suited manner. Currently efficient modelling tools have been developed for different length scales and there still exists a challenge in passing relevant information between models on different scales. This chapter aims at overviewing the current stage of modelling tools at different length scales, discussing the possible approaches to bridge different length scales, and reporting successful multiscale modelling applications.

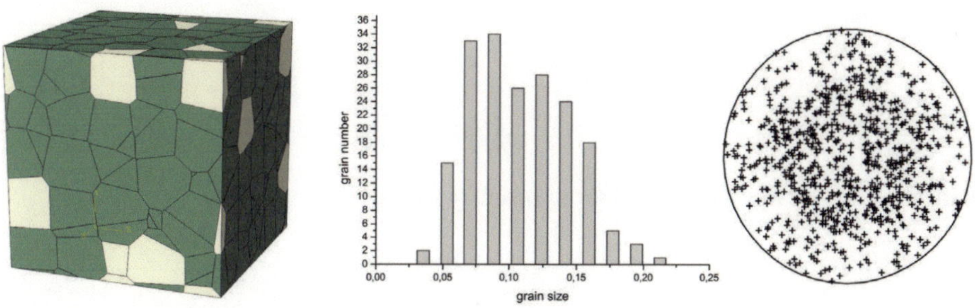

Fig. 1. Multiphase polycrystalline RVE (right) with 90% matrix and 10% precipitate. The grain size has a normal distribution (middle) and the [111] polfigure (left) shows a random texture.

2. Generating realistic material microstructures

The current advanced high strength steels (AHSS) such as dual phase steels, transformation induced plasticity (TRIP) steels, twin induced plasticity (TWIP) steels and martensitic steels are all multiphase polycrystalline materials. In order to model the the macroscopic mechanical properties such as yield stress, work hardening rate and elongation to fracture, one has to build a representative volume element (RVE) for each macroscale material point and investigate the local deformation of each material point within the RVE, and then make a volume average. In this micro-macro-transition procedure, in order to reduce the computational costs the statistically similar representative volume elements (SSRVEs) have been developed to replace real microstructures from metallurgical images by Schröder et al. (2010).

Considering the real microstructure of multiphase materials, during the representative volume element generation one should consider grain shape distribution, crystalline orientation distribution, grain boundary misorientation angle distribution and volume fraction of different phases. Figure 1 is an example of the RVE we have generated for TRIP steels where the Voronoi tessellation algorithm has been used.

Recent studies (Lu et al., 2009; 2004) show bulk specimens comprising nanometer sized grains with embedded lamella with coherent, thermal and mechanical stable twin boundaries exhibiting high strength and considerable ductility at the same time. These materials have higher loading rate sensitivity, better tolerance to fatigue crack initiation, and greater resistance to deformation. Under this condition, the RVE with nanometer sized twin lamella inside nanometer sized twin lamella inside nanometer sized grains will help us to understand existing material behavior and design new materials.

Assume two orientations $\mathbf{Q}_\mathrm{I}$ and $\mathbf{Q}_\mathrm{II}$ have the twin relationship in $(1,1,1)$ habit plane along $[1,1,-2]$ twinning direction. For any vector $\mathbf{V}$, these two tensors will map as $\mathbf{v}_\mathrm{I} = \mathbf{Q}_\mathrm{I}\mathbf{V}$ and $\mathbf{v}_\mathrm{II} = \mathbf{Q}_\mathrm{II}\mathbf{V}$. The twin relationship between $\mathbf{v}_\mathrm{I}$ and $\mathbf{v}_\mathrm{II}$ is easier to see in the local twin coordinate system with $\mathbf{x}'//[1,1,-2]$ and $\mathbf{z}'//[1,1,1]$ rather than in the global coordinate system $[\mathbf{x},\mathbf{y},\mathbf{z}]$. We define a orthogonal tensor $\mathbf{R}_\mathrm{L}$ for the mapping from global coordinate system to the local coordinate system

$$R_{\mathrm{L}ij} = \begin{bmatrix} \frac{1}{\sqrt{6}} & \frac{1}{\sqrt{2}} & \frac{1}{\sqrt{3}} \\ \frac{1}{\sqrt{6}} & \frac{-1}{\sqrt{2}} & \frac{1}{\sqrt{3}} \\ \frac{-2}{\sqrt{6}} & 0 & \frac{1}{\sqrt{3}} \end{bmatrix} \tag{1}$$

and another tensor $\mathbf{R}_\mathrm{M}$ for the mirror symmetry operation

$$R_{\mathrm{M}ij} = \begin{bmatrix} 1 & 0 & 0 \\ 0 & 1 & 0 \\ 0 & 0 & -1 \end{bmatrix} \tag{2}$$

and get the twin relationship in the local twin coordinate system

$$\left(\mathbf{R}_\mathrm{L}\mathbf{Q}_\mathrm{I}\mathbf{R}_\mathrm{L}^\mathrm{T}\right)\mathbf{v}' = \mathbf{R}_\mathrm{M}\left(\mathbf{R}_\mathrm{L}\mathbf{Q}_\mathrm{II}\mathbf{R}_\mathrm{L}^\mathrm{T}\right)\mathbf{v}'. \tag{3}$$

Because $\mathbf{v}'$ is a arbitrary vector in the $[\mathbf{x}',\mathbf{y}',\mathbf{z}']$ coordinate system equation 3 will reduce to

$$\mathbf{R}_{\mathrm{L}}\mathbf{Q}_{\mathrm{I}}\mathbf{R}_{\mathrm{L}}^{\mathrm{T}} = \mathbf{R}_{\mathrm{M}}\left(\mathbf{R}_{\mathrm{L}}\mathbf{Q}_{\mathrm{II}}\mathbf{R}_{\mathrm{L}}^{\mathrm{T}}\right) \tag{4}$$

and we find the relationship between $\mathbf{Q}_{\mathrm{I}}$ and $\mathbf{Q}_{\mathrm{II}}$ as

$$\mathbf{Q}_{\mathrm{II}} = \mathbf{R}_{\mathrm{L}}^{\mathrm{T}}\mathbf{R}_{\mathrm{M}}^{-1}\mathbf{R}_{\mathrm{L}}\mathbf{Q}_{\mathrm{I}}. \tag{5}$$

The crystal orientation $\mathbf{Q}_{\mathrm{I}}$ has been assigned to the material point at the centre of the individual grain with coordinate $[\mathbf{x}_0',\mathbf{y}_0',\mathbf{z}_0']$. The orientation $\mathbf{Q}_{\mathrm{II}}$ will be assigned to the twinned material point with coordinate $[\mathbf{x}_1',\mathbf{y}_1',\mathbf{z}_1']$ when the distance between this point and the grain center along the habit plane normal direction and the lamella thickness d satisfy

$$\left[(2k-1)-\frac{1}{2}\right]\cdot d < |z_1'-z_0'| \leq \left[2k-\frac{1}{2}\right]\cdot d \text{ with } k=1,2,3,... \tag{6}$$

Otherwise orientation $\mathbf{Q}_{\mathrm{I}}$ will be assigned to this material point.

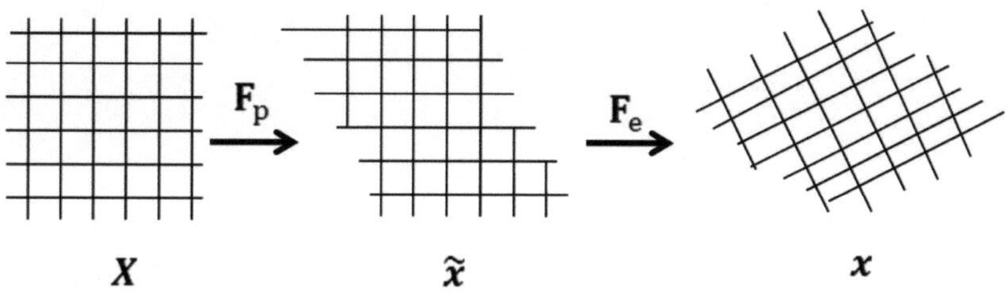

Fig. 2. The multiplicative decomposition of the deformation gradient where the plastic deformation is accommodated by dislocation slip.

3. Constitutive models based on continuum mechanics

3.1 Kinematics

For large strains the elastic and plastic deformation can be separated consistently. We follow the well-known multiplicative decomposition proposed by Lee (1969) (see Figure 2) of the deformation gradient tensor $\mathbf{F}$

$$\mathbf{F} = \frac{\partial \mathbf{x}}{\partial \mathbf{X}} = \frac{\partial \mathbf{x}}{\partial \tilde{\mathbf{x}}}\frac{\partial \tilde{\mathbf{x}}}{\partial \mathbf{X}} = \mathbf{F}_{\mathrm{e}}\mathbf{F}_{\mathrm{p}} \tag{7}$$

where $\mathbf{F}_{\mathrm{e}}$ is the elastic part comprising the stretch and rotation of the lattice, and $\mathbf{F}_{\mathrm{p}}$ corresponds to the plastic deformation. The lattice rotation $\mathbf{R}_{\mathrm{e}}$ and stretch $\mathbf{U}_{\mathrm{e}}$ are included in the mapping $\mathbf{F}_{\mathrm{e}}$. They can be calculated by the polar decomposition $\mathbf{F}_{\mathrm{e}} = \mathbf{R}_{\mathrm{e}}\mathbf{U}_{\mathrm{e}}$, i.e., the texture evolution is included in this part of the deformation. Furthermore, two rate equations can be derived for the elastic and the plastic deformation gradients as

$$\dot{\mathbf{F}}_{\mathrm{e}} = \mathbf{L}\mathbf{F}_{\mathrm{e}} - \mathbf{F}_{\mathrm{e}}\mathbf{L}_{\mathrm{p}} \tag{8}$$

$$\dot{\mathbf{F}}_{\mathrm{p}} = \mathbf{L}_{\mathrm{p}}\mathbf{F}_{\mathrm{p}} \tag{9}$$

where $\mathbf{L} = \dot{\mathbf{F}}\mathbf{F}^{-1}$ and $\mathbf{L}_\mathrm{p} = \dot{\mathbf{F}}_\mathrm{p}\mathbf{F}_\mathrm{p}^{-1}$ are the total and the plastic velocity gradients defined in the current and the unloaded configuration respectively. Because the stress produced by the elastic deformation can supply driving forces for dislocation slip, twinning formation and phase transformation which can accommodate the plastic deformation, $\mathbf{F}_\mathrm{e}$ and $\mathbf{F}_\mathrm{p}$ are not independent. If the total deformation process is known, the elastic and plastic deformation evolutions can be determined through solving equations 7,8 and 9.

When the eigen-deformation $\mathbf{F}_\mathrm{t}$ of phase transformation and the plastic deformation $\mathbf{F}_\mathrm{p}$ of dislocation slip coexist, the multiply decomposition (see Figure 3) should be reformulated as the following

$$\mathbf{F} = \frac{\partial \mathbf{x}}{\partial \widetilde{\mathbf{x}}'}\frac{\partial \widetilde{\mathbf{x}}'}{\partial \widetilde{\mathbf{x}}}\frac{\partial \widetilde{\mathbf{x}}}{\partial \mathbf{X}} = \mathbf{F}_\mathrm{e}\mathbf{F}_\mathrm{t}\mathbf{F}_\mathrm{p}. \tag{10}$$

The evolution of $\mathbf{F}_\mathrm{t}$ is controlled by the transformed volume fraction f_α because the eigen-deformation $\widetilde{\mathbf{H}}_\mathrm{t}^\alpha$ of each transformation system with unit volume fraction is a constant tensor

$$\mathbf{F}_\mathrm{t} = \mathbf{I} + \sum_{\alpha=1}^{N_T} f^\alpha \widetilde{\mathbf{H}}_\mathrm{t}^\alpha \tag{11}$$

$$\dot{\mathbf{F}}_\mathrm{t} = \sum_{\alpha=1}^{N_T} \dot{f}^\alpha \widetilde{\mathbf{H}}_\mathrm{t}^\alpha. \tag{12}$$

where N_T is the total number of transformation system.

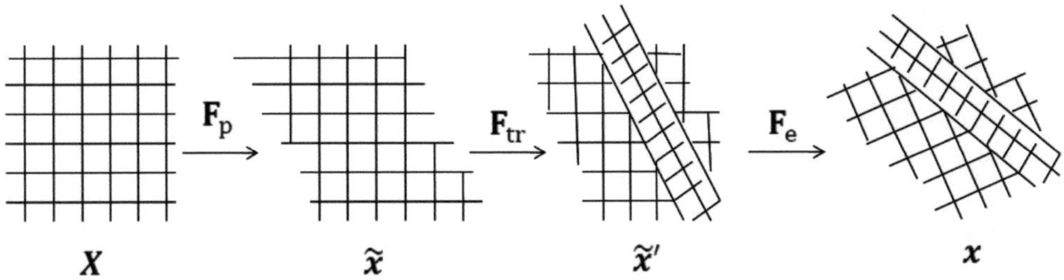

Fig. 3. The multiplicative decomposition of the deformation gradient when dislocation slip and phase transformation coexist.

3.2 The elastic deformation

For the dislocation slip case, the plastic deformation $\mathbf{F}_\mathrm{p}$ will not change the lattice orientation, i.e., we can use a constant stiffness tensor $\widetilde{\mathbb{K}}_0$ for the stress calculations and define the elastic law in the unloaded configuration. By defining the second Piola-Kirchhoff stress tensor $\widetilde{\mathbf{S}}$ in the unloaded configuration and its work conjugated elastic Green strain tensor $\tilde{\mathbf{E}}$, the elastic law is derived as

$$\widetilde{\mathbf{S}} = \widetilde{\mathbb{K}}_0\widetilde{\mathbf{E}} \tag{13}$$

with

$$\widetilde{\mathbf{E}} = \frac{1}{2}\left(\mathbf{F}_\mathrm{e}^\mathrm{T}\mathbf{F}_\mathrm{e} - \mathbf{I}\right) \tag{14}$$

where $\mathbf{I}$ is the second order unity tensor. When the variational principle of the FEM is formulated in the reference or current configuration, the second Piola-Kirchhoff stress $\mathbf{S}$ or the Cauchy stress $\boldsymbol{\sigma}$ amount to

$$\mathbf{S} = \mathbf{F}_p^{-1}\,\widetilde{\mathbf{S}}\,\mathbf{F}_p^{-T} \tag{15}$$

$$\boldsymbol{\sigma} = \frac{1}{J}\mathbf{F}_e\,\widetilde{\mathbf{S}}\,\mathbf{F}_e^T \tag{16}$$

where $J = det(\mathbf{F}) = det(\mathbf{F}_e)$, which means the isochoric plastic deformation is assumed, i.e., the volume change is always purely elastic.

In a polycrystal the different grains have different initial orientations. Therefore, in the global coordinate system, different stiffness tensors, slip plane normals and slip directions should be specified for every crystal. In order to use only one set of data, a virtual deformation step is introduced before the calculation in the following form: $\mathbf{F}_{p0}$ is set as the initial value for $\mathbf{F}_p$. By choosing $\mathbf{F}_{e0}$ to satisfy

$$\mathbf{F}_{e0}\,\mathbf{F}_{p0} = \mathbf{I}, \quad \mathbf{F}_{e0}, \mathbf{F}_{p0} \in Orth. \tag{17}$$

the starting value for $\mathbf{F}$ amounts to $\mathbf{I}$ as desired. If one adopts the Bunge Euler angle $(\varphi_1, \Phi, \varphi_2)$ to define the crystal orientation, the matrix of the elastic deformation gradient amounts to

$$F_{e0} = \begin{bmatrix} \cos(\varphi_2) & -\sin(\varphi_2) & 0 \\ \sin(\varphi_2) & \cos(\varphi_2) & 0 \\ 0 & 0 & 1 \end{bmatrix} \begin{bmatrix} 1 & 0 & 0 \\ 0 & \cos(\Phi) & -\sin(\Phi) \\ 0 & \sin(\Phi) & \cos(\Phi) \end{bmatrix} \begin{bmatrix} \cos(\varphi_1) & -\sin(\varphi_1) & 0 \\ \sin(\varphi_1) & \cos(\varphi_1) & 0 \\ 0 & 0 & 1 \end{bmatrix}.$$

3.3 The dislocation slip based plastic deformation

The plastic deformation mechanism discussed here is the slip mechanism where dislocations slip in certain crystallographic planes along certain crystallographic directions to accommodate shape changes of the crystal. This is the most common mechanism in conventional metal forming processes.

The concept for describing displacement fields around dislocations in crystals was developed mathematically by Volterra and used for calculating elastic deformation fields by Orowan and Taylor in 1934 (Hirth & Lothe, 1992). Dislocations are one dimensional lattice defects which can not begin or end inside a crystal, but must intersect a free surface, form a closed loop or make junctions with other dislocations. Due to energetic reasons there is a strong tendency for dislocations to assume a minimum Burgers vector, and to slide in the planes with maximum interplanar separation and along the most densely packed directions. Under the applied stress, the lattice deforms elastically, until stretched bonds near the dislocation core break and new bonds form. The dislocation moves step by step by one Burgers vector. It is the dislocation slip mechanism that can explain why the actually observed strength of most crystalline materials is between one to four orders of magnitude smaller than the intrinsic or theoretical strength required for breaking the atom bonds without the presence of dislocations.

In order to set the connection between the continuum variables and the process of dislocation slip, we have to determine the shear amount of individual slip systems. The slip systems are mathematically described by the Schmid tensor $\widetilde{\mathbf{M}}^\alpha = \widetilde{\mathbf{d}}^\alpha \otimes \widetilde{\mathbf{n}}^\alpha$ where $\widetilde{\mathbf{d}}^\alpha = \mathbf{b}/b$ expresses the slip direction, which is parallel to the Burgers vector $\mathbf{b}$, but normalised, and $\widetilde{\mathbf{n}}^\alpha$, the slip plane normal with respect to the undistorted configuration. Through calculating the line vector $\widetilde{\mathbf{l}}^\alpha = \widetilde{\mathbf{d}}^\alpha \times \widetilde{\mathbf{n}}^\alpha$ we can define one local coordinate system for slip system α as $[\widetilde{\mathbf{l}}^\alpha, \widetilde{\mathbf{d}}^\alpha, \widetilde{\mathbf{n}}^\alpha]$, which is useful in the later GND calculations.

For the FCC crystal structure, the close-packed planes $\{111\}$ and close-packed directions $\langle 110\rangle$ form 12 slip systems. For the BCC crystal structure, the pencil glide phenomenon is observed, which resembles slip in a fixed direction on apparently random planes. In literature, experimental studies have shown that for BCC crystals the slip direction is along $\langle 111\rangle$, and three groups of slip planes exist, including $\{110\}$,$\{112\}$ and the less common $\{123\}$ planes. Totally there are 48 slip systems for BCC crystals. Therefore, for FCC and BCC crystals it is possible to supply five independent slip systems to accommodate any arbitrary external plastic deformation, and in the middle temperature range the slip is the main mechanism for plastic deformations. For the HCP crystal, the slip system number is dependent on the axis ratio of the HCP unit cell. When this ratio assumes values in a certain range as discussed by Gottstein (2004), only two independent slip systems exist, and it is impossible to accommodate a arbitrary deformation by slip steps. As a result, mechanical twins appear during plastic deformation.

Among all of the dislocations in one slip system, only the mobile dislocations can produce plastic deformation, and their speed can be determined by the forces acting on them. In general, the driving force is related to external loads, short range isotropic resistance of dislocation interactions and long range back stress of dislocation pile-ups and lattice frictions. The widely-adopted constitutive assumption for crystal plasticity reads

$$\mathbf{L}_\mathrm{p} = \sum_{\alpha=1}^{N_S} \dot{\gamma}^\alpha \widetilde{\mathbf{M}}^\alpha \,, \qquad \dot{\mathbf{F}}_\mathrm{p} = \sum_{\alpha=1}^{N_S} \dot{\gamma}^\alpha \widetilde{\mathbf{M}}^\alpha \mathbf{F}_\mathrm{p} \tag{18}$$

where $\dot{\gamma}^\alpha$ is the slip rate on slip system α within the intermediate configuration $\widetilde{\mathbf{x}}$, and N_S, the total number of slip systems.

When a part of material is transferred into another lattice with volume fraction $f = \sum_{\alpha=1}^{N_T} f^\alpha$ which can only deform elastically, equation 18 has been modified as

$$\mathbf{L}_\mathrm{p} = (1-f) \sum_{\alpha=1}^{N_S} \dot{\gamma}^\alpha \widetilde{\mathbf{M}}^\alpha \,, \qquad \dot{\mathbf{F}}_\mathrm{p} = (1-f) \sum_{\alpha=1}^{N_S} \dot{\gamma}^\alpha \widetilde{\mathbf{M}}^\alpha \mathbf{F}_\mathrm{p}. \tag{19}$$

3.3.1 The Orowan equation

Commonly used expressions for the relation of the shear rate, $\dot{\gamma}$, and the resolved shear stress, τ, include a phenomenological viscoplastic law in the form of a power law by Peirce et al. (1982), and more physically-based ones such as those of Kocks et al. (1975) and Nemat-Nasser et al. (1998), which can take account of rate and temperature dependencies. In this paper we use the Orowan equation to calculate the plastic shear rate $\dot{\gamma}$ of each slip system as a function of the mobile dislocation density, ρ_m, on that slip system

$$\dot{\gamma} = \rho_\mathrm{m}\, b\, v \tag{20}$$

where the average velocity of the mobile dislocations, v, is a function of the resolved shear stress, τ, of the dislocation densities, ρ_M, ρ_SSD and ρ_GND and its gradient, the average GND pile-up size, L, and of the temperature, θ; i.e.,

$$v = v\left(\tau, \rho_{\text{SSD}}, \rho_{\text{GND}}, \frac{\partial \rho_{\text{GND}}}{\partial X}, L, \theta\right) \tag{21}$$

The resolved shear stress, τ, is the projection of the stress measure onto the slip system. In the case of infinitesimally small elastic stretches $\left\|\widetilde{\mathbf{C}}\right\| = \left\|\widetilde{\mathbf{F}}_{\text{e}}^{\text{T}}\widetilde{\mathbf{F}}_{\text{e}}\right\| << 1$, the resolved shear stress, τ, within the intermediate configuration $\widetilde{\mathbf{x}}$ can be approximated by following Kalidindi et al. (1992)

$$\tau^{\alpha} = \widetilde{\mathbf{S}}\,\widetilde{\mathbf{C}} \cdot \widetilde{\mathbf{M}^{\alpha}} \cong \tilde{\mathbf{S}} \cdot \widetilde{\mathbf{M}^{\alpha}} \tag{22}$$

In order to accommodate a part of the external plastic deformation, the mobile dislocations, ρ_{M}, must overcome the stress field of the parallel dislocations, ρ_{P}, which cause the passing stress. They must also cut the forest dislocations, ρ_{F}, with the aid of thermal activation. We define the parallel dislocation density and the forest dislocation density as: ρ_{P} for all dislocations parallel to the slip plane, and ρ_{F} for the dislocations perpendicular to the slip plane. Both ρ_{SSD} and ρ_{GND} are contributing to ρ_{F} and ρ_{P}

$$\rho_{\text{F}}^{\alpha} = \sum_{\beta=1}^{N_S} \chi^{\alpha\beta}\left(\rho_{\text{SSD}}^{\beta} + \rho_{\text{GND}}^{\beta}\right)\left|\cos(\widetilde{\mathbf{n}}^{\alpha}, \widetilde{\mathbf{t}}^{\beta})\right| \tag{23}$$

and

$$\rho_{\text{P}}^{\alpha} = \sum_{\beta=1}^{N_S} \chi^{\alpha\beta}\left(\rho_{\text{SSD}}^{\beta} + \rho_{\text{GND}}^{\beta}\right)\left|\sin(\widetilde{\mathbf{n}}^{\alpha}, \widetilde{\mathbf{t}}^{\beta})\right| \tag{24}$$

where we introduce the interaction strength, $\chi^{\alpha\beta}$, between different slip systems, which includes the self interaction strength, coplanar interaction strength, cross slip strength, glissile junction strength, Hirth lock strength, and Lomer-Cottrell lock strength. One can go further to see the definition of these interactions in literature (Devincre et al., 2008; Madec et al., 2008). In this formulation we only consider edge dislocations owing to their limited mobility for the FCC crystal, and use a single set of interaction strengths for both SSDs and GNDs.

With the help of the forest dislocation density ρ_{F}, we can determine the average jump distance of the mobile dislocation and the activation volume for the thermal activated forest dislocation cutting event

$$\lambda = \frac{c_1}{\sqrt{\rho_F}} \tag{25}$$

and

$$V = c_2 b^2 \lambda \tag{26}$$

where c_1 and c_2 are constants to reflect the real dislocation line configuration which is more complicated than the schematic pictures we use here.

With the help of the parallel dislocation density ρ_{P} and the gradient of GND density $\frac{\partial \rho_{\text{GND}}}{\partial X}$, we can determine the average athermal passing stress τ_{p} and back stress τ_{b} as following

$$\tau_{\text{p}} = c_3 G b \sqrt{\rho_P} \tag{27}$$

and

$$\tau_{\mathrm{b}} = GbL^2 \frac{\partial \rho_{\mathrm{GND}}}{\partial X} \tag{28}$$

where c_3 is the constant for the Taylor hardening mechanism. For reasons of simplicity, in equation 28 the back stress of one slip system only comes from the GND pile-up of this slip system. This equation can be easily extended to consider the back stress from all of the slip system at the same time.

Compared with flow rules in the literature which contain a constant reference shear rate and a constant rate sensitivity exponent, here a flow rule is derived based on the dislocation slip mechanism

$$\dot{\gamma} = \begin{cases} \rho_{\mathrm{m}} b \lambda \nu_0 \exp\left(-\frac{Q_{\mathrm{slip}}}{\mathrm{k_B}\theta}\right) \exp\left(\frac{|\tau+\tau_{\mathrm{b}}|-\tau_{\mathrm{p}}}{\mathrm{k_B}\theta} V\right) \mathrm{sign}\,(\tau+\tau_{\mathrm{b}}) & |\tau+\tau_{\mathrm{b}}| > \tau_{\mathrm{p}} \\ 0 & |\tau+\tau_{\mathrm{b}}| \leq \tau_{\mathrm{p}} \end{cases} \tag{29}$$

where $\mathrm{k_B}$ is the Boltzmann constant, ν_0 the attempt frequency and Q_{slip} the effective activation energy.

Inside the flow rule given by equation 29 determination of the mobile dislocation density is a hard task. In some research work, the mobile dislocation density was found to be a small fraction of total dislocation density and is even treated as a constant. The more sophisticated model to deal with this dislocation density based on energy minimization can be found in Ma & Roters (2004). For reasons of simplicity here the mobile dislocation density is treated as a constant number.

3.3.2 Evolution of the dislocation densities

There are four processes contributing to the evolution of the SSD density as discussed by Ma (2006). The lock forming mechanism between mobile dislocations and forest dislocations, the dipole forming mechanism between mobile dislocations with parallel line vectors, and anti-parallel Burgers vector determine the multiplication terms, while the annihilation term includes annihilation between one mobile dislocation with another immobile one and annihilation between two immobile dislocations. The often used Kocks-Mecking model, as discussed in Roters (1999), only adopts the locks formation and mobile-immobile annihilation mechanisms for the SSD evolution

$$\dot{\rho}_{\mathrm{SSD}} = \left(c_4 \sqrt{\rho_{\mathrm{F}}} - c_5\, \rho_{\mathrm{SSD}}\right) \dot{\gamma} \tag{30}$$

Where c_4 and c_5 are constants used to adjust the locks and annihilation radius.

When plastic deformation gradients are present in a volume portion, GNDs must be introduced to preserve the continuity of the crystal lattice. A relation between a possible GND measure and the plastic deformation gradient has been proposed by Nye (1953). This approach has been later extended to a more physically motivated continuum approach to generally account for strain gradient effects by Dai & Parks (1997). Following these pioneering approaches, we use as a dislocation density tensor, $\mathbf{\Lambda}$, for a selected volume portion to calculate the net Burgers vector for an area

$$\boldsymbol{\Lambda} = \bar{\mathbf{b}} \otimes \bar{\mathbf{l}} = -\left(\boldsymbol{\nabla}_{\mathrm{X}} \times \mathbf{F}_{\mathrm{p}}\right)^{\mathrm{T}} \tag{31}$$

where $\boldsymbol{\nabla}_{\mathrm{X}} = \partial/\partial\mathbf{X}$, is defined as the derivative with respect to the reference coordinates and $\bar{\mathbf{b}}$ and $\bar{\mathbf{l}}$ are, respectively, the net Burgers vector and net line vector after an volume average operation. Using equation (31) the resulting Burgers vector for a circuit with an arbitrary orientation can be calculated. In general this tensor is non-symmetric and it can be mapped to nine independent slip systems in a unique fashion. For the FCC crystal structure with its 12 slip systems, only 5 systems are independent according to the von Mises-Taylor constraint. This implies that it is impossible to calculate the exact amount of GNDs for every slip system in a unique way. Nevertheless, we can project $\boldsymbol{\Lambda}$ to each of the slip systems to determine the Burgers vector of the edge and screw type GNDs for the pass stress and backing stress calculation

$$\mathbf{b}^{\alpha}_{\mathrm{GNDe}} = \left(\widetilde{\mathbf{d}}^{\alpha} \cdot \boldsymbol{\Lambda} \cdot \widetilde{\mathbf{l}}^{\alpha}\right) \widetilde{\mathbf{d}}^{\alpha} \tag{32}$$

and

$$\mathbf{b}^{\alpha}_{\mathrm{GNDs}} = \left(\widetilde{\mathbf{d}}^{\alpha} \cdot \boldsymbol{\Lambda} \cdot \widetilde{\mathbf{d}}^{\alpha}\right) \widetilde{\mathbf{d}}^{\alpha}. \tag{33}$$

Furthermore we also can calculate the GND density as the following

$$\rho^{\alpha}_{\mathrm{GND}} = \left(\|\mathbf{b}^{\alpha}_{\mathrm{GNDe}}\| + |\mathbf{b}^{\alpha}_{\mathrm{GNDs}}|\right)/b. \tag{34}$$

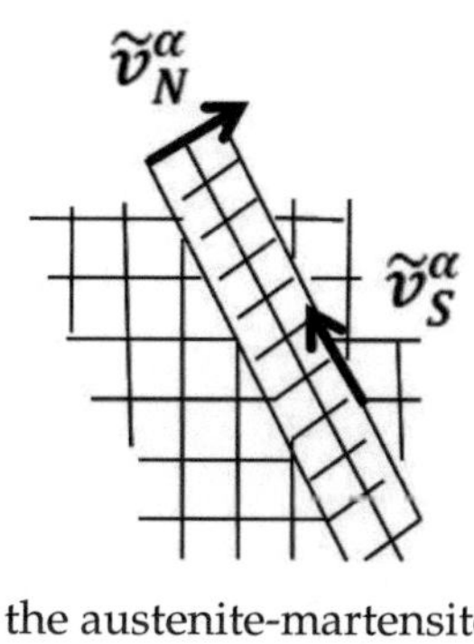

Fig. 4. A transformation system of the austenite-martensite phase transformation.

3.4 Eigenstrain of phase transformations

The transformation-induced plasticity (TRIP) assisted steels are mixtures of allotriomorphic ferrite, bainite and retained austenite. Experimental and modelling publications have highlighted that the transformation of retained austenite to martensite under the influence of a applied stress or strain can improve material ductility and strength efficiently, as shown by Bhadeshia (2002).

According to the geometrically nonlinear theory of martensitic transformations (Bhattacharya, 1993; Hane & Shield, 1998; 1999) there are 24 transformation systems and they are constructed by two body-centered tetragonal (BCT) variants with relative rotations and volume fractions, in order to produce habit planes between austenite and martensite arrays and pairwise arranging twin related variant lamellas. Each transformation system corresponds to one constant shape strain vector, $\widetilde{\mathbf{v}}_s^\alpha$, and one constant habit plane normal vector, $\widetilde{\mathbf{v}}_n^\alpha$, see Figure 4. Following the classical Kurdjumov-Kaminsky relations, these two vectors are influenced by the carbon concentration through the lattice parameter magnitude variation (Hane & Shield, 1998; Wechsler et al., 1953). The eigenstrain of the transformation system α amounts to

$$\widetilde{\mathbf{H}}_\mathrm{t}^\alpha = \widetilde{\mathbf{v}}_s^\alpha \otimes \widetilde{\mathbf{v}}_n^\alpha \tag{35}$$

As an example in the literature (Kouznetsova & Geers, 2008; Tjahjanto, 2008), the shape strain vector and the habit plane normal vector with respect to specific carbon concentrations have been determined and listed. Because the shape strain vector and the habit plane normal vector are explicit functions of austensite and martensite lattice parameters, the components of the two vectors are irrational. $\widetilde{\mathbf{H}}_\mathrm{t}^\alpha$ is similar to the Schmid tensor $\widetilde{\mathbf{M}}^\alpha$ of dislocation slip except that there is volume change $det(\widetilde{\mathbf{H}}_\mathrm{t}^\alpha) > 1$.

In literature, the austenite-martensite phase transformation has been formulated as stress and strain driving mechanisms among different temperature regions. The stress controlled transformation often occurs at lower temperatures where the chemical driving force is so high that a external load below austenite yield stress can help the already existing martensite nuclei to grow. At strain controlled transformation regions at higher temperatures, the chemical driving force is so low that additional loads which are higher than the yield stress are needed in order for existing nuclei to continue grow. Due to the fact that plastic deformation in austenite is easier than spontaneous martensite formation, the transformation has to be continued by new nuclei formation at the shear band intersection region according to Olson & Cohen (1972).

Following the Olson-Cohen model, the transformation kinetics are formulated in meso-scale on the micro-band level. At first the shear band density is estimated, then the intersection frequency of shear bands is calculated and lastly the nucleation producing probability is evaluated. This governing equation for martensite volume fraction reads

$$\dot{f}^\alpha = \begin{cases} c_6(1 - \sum_{\beta=1}^{N_\mathrm{T}} f_\beta)\, |\tau_\alpha|^{c_7} \sum_{i,j=1}^{N_\mathrm{S}} \left(1 - \left|\widetilde{\mathbf{l}}_i \cdot \widetilde{\mathbf{l}}_j\right|\right) \left(1 - \left|\widetilde{\mathbf{n}}_i \cdot \widetilde{\mathbf{n}}_j\right|\right) \sqrt{\left|\dot{\gamma}_i \dot{\gamma}_j\right|} & \tau_\alpha > 0 \\ 0 & \tau_\alpha \le 0 \end{cases} \tag{36}$$

where N_T and N_S are the total number of transformation systems and slip systems with, respectively, c_6 and c_7 as two fitting parameters to control the transformation kinetics; and

$$\tau^\alpha \cong \tilde{\mathbf{S}} \cdot \widetilde{\mathbf{H}}_\mathrm{t}^\alpha \tag{37}$$

the resolved stress in the transformation system α with includes the shear part and the tensile or compression part at the same time. Indeed, in equation 36 the phase transformation is controlled by the external load potential minimisation.

Because the transformed martensitic phase includes twinned wedge microstructures, the Frank-Read dislocation source may suffer higher resistance compared with the original austenitic phase. The dislocation slip based plasticity of martensite has been neglected here.

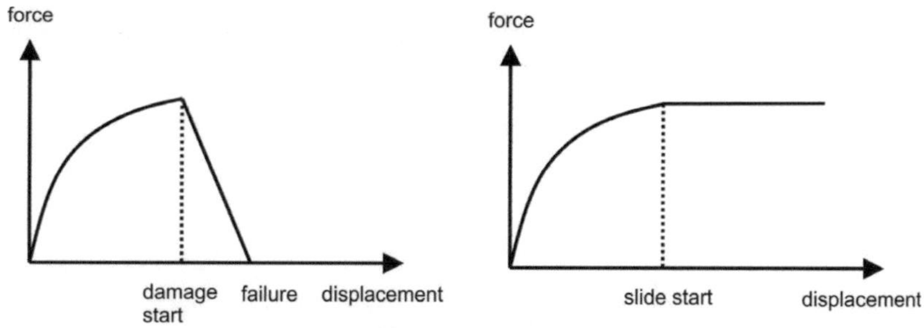

Fig. 5. Schematic drawings of cohesive behaviour of grain boundaries along normal (left) and tangential (right) directions.

3.5 Cohesive zone model for grain and phase boundaries

In experimental works and atomistic simulations with respect to deformation of nanocrystalline materials, dislocation glide inside the grains and grain boundary sliding have both been reported. It is obvious that grain-boundary sliding and separation mechanisms begin to play important roles in the overall inelastic response of a polycrystalline material when the grain-size decreases and dislocation activity within the grain interior becomes more difficult. In recent work, the atomic bonds across grain boundaries have been characterized with ab initio calculations within the framework of the density functional theory (Janisch et al., 2010). In this work not only the energetics of grain boundaries have been characterized, but also the mechanical response of a grain boundary to applied loads is studied. Such information can be used to parameterize cohesive zone models based on ab initio calculations.

The cohesive zone model is useful for RVE models of polycrystals, in situations when grain boundary deformation needs to be taken into account explicitely, e.g. when grain boundary sliding or damage initiation at grain boundaries or phase bounaries has to be considered. By adjusting the cohesive zone parameters for grain boundary sliding and opening the competing mechanisms of bulk material deformation and grain boundary accommodated deformation can be studied. Furthermore, it is also possible to investigate damage nucleation at GB triple junctions.

We follow Wei & Anand (2004) to generate a rate independent cohesive zone (CZ) modelling approach for the reasons of simplicity. The velocity jump across a cohesive surface has been additively decomposed into a elastic and a plastic part as follow

$$\dot{\mathbf{u}} = \dot{\mathbf{u}}_{\mathrm{e}} + \dot{\mathbf{u}}_{\mathrm{p}}. \tag{38}$$

The elastic relative velocities are connected with its power-conjugate traction rate by the interface elastic stiffness tensor

$$\dot{\mathbf{t}} = \mathbf{K}\dot{\mathbf{u}}_{\mathrm{e}} = \mathbf{K}\left(\dot{\mathbf{u}} - \dot{\mathbf{u}}_{\mathrm{p}}\right) \tag{39}$$

For some special grain boundaries there may exist glide anisotropy inside the grain boundary plane, although our knowledge about this topic is far away from formulating this anisotropy for general grain boundaries. So, we have to assume isotropic plastic deformation property inside the grain boundary, and the trace vector, displacement vector and resistance vector are defined in the local coordinate $[\mathbf{t}_I, \mathbf{t}_{II}, \mathbf{n}]$, where $\mathbf{n}$ is aligned with the normal to the interface, $\mathbf{t}_I$ and $\mathbf{t}_{II}$ in the tangent plane at the point of the interface under consideration.

According to the rate independent assumption for the loading condition, the hardening rate $\dot{\mathbf{s}}$ has to be fast enough to balance the load $\dot{\mathbf{t}}$

$$\dot{\mathbf{t}} = \dot{\mathbf{s}} \tag{40}$$

or

$$\mathbf{K}\left(\dot{\mathbf{u}} - \dot{\mathbf{u}}_{\mathrm{p}}\right) = \mathbf{H}\dot{\mathbf{u}}_{\mathrm{p}} \tag{41}$$

where $\mathbf{H}$ is the hardening matrix and its components are state variables of the cohesive zone model. The evolution law of the hardening matrix

$$\dot{\mathbf{H}} = \dot{\mathbf{H}}\left(\mathbf{H}, \dot{\mathbf{u}}_{\mathrm{p}}\right) \tag{42}$$

can be obtained by experimental date fitting or can directly come from the molecular dynamic and ab-initial calculations. Here we adopted the phenomenological hardening rule proposed by Wei & Anand (2004) for the cohesive zone model along the normal direction. For the cohesive zone model along the tangential direction, the failure displacement has been set to infinite to consider the grain boundary glide phenomenon, see Figure 5.

4. Numerical approaches

Starting from the stress equilibrium state, for a given time step and velocity gradient one has to calculate the new stress state, while at the same time considering the evolution of state variables including plastic shear amount γ determined by equation 29, statistically stored dislocation density ρ_{SSD} determined by equation 30 and geometrically necessary dislocation density ρ_{GND} determined by equation 34 for each slip system, the transformed volume fraction f determined by equation 36 for each transformation system and the hardening matrix $\mathbf{H}$ determined by equation 42 for the cohesive zone model of grain boundaries.

4.1 Finite element method

Based on the Abaqus platform (ABAQUS, 2009), we have developed the user material subroutines UMAT for the bulk material and UINTER for the grain boundary to solve the stress equilibrium and state variable evolution problems. In this approach, except for the plastic strain gradient used for the geometrically necessary dislocation density calculation, which is adopted from last converged time point, all of state variables are calculated by an implicit method.

4.2 Discrete fast Fourier transformation method

If the representative volume element has a very complicated microstructure and obeys a periodic boundary condition, the stress equilibrium and state variable evolution problems can be solved by the discrete fast Fourier transformation (FFT) method proposed in the literature (Lebensohn, 2001; Michel et al., 2000; 2001).

According to this approach the material points of the real RVE are approximated to inclusions inside one homogeneous matrix, the property of which can be determined as volume average of these inclusions. After the regular discretisation of the RVE, in the current configuration the inclusion at location $\mathbf{x}'$ has stress increment $\Delta\boldsymbol{\sigma}^I$, strain increment $\Delta\boldsymbol{\epsilon}^I$ and stiffness $\mathbb{C}^I = \partial\Delta\boldsymbol{\sigma}^I/\partial\Delta\boldsymbol{\epsilon}^I$, and the matrix material point at the same location has stress increment $\Delta\boldsymbol{\sigma}^M$, strain increment $\Delta\boldsymbol{\epsilon}^M$ and stiffness $\bar{\mathbb{C}}$.

Assuming we deal with a deformation control process, at the beginning of the iteration loop

$$\Delta\boldsymbol{\epsilon}^I = \Delta\boldsymbol{\epsilon}^M = \overline{\Delta\boldsymbol{\epsilon}} \tag{43}$$

where $\overline{\Delta\boldsymbol{\epsilon}}$ is the given fixed strain increment. Because each material point has the same volume and shape, the matrix stiffness can be determined simply as

$$\bar{\mathbb{C}} = \sum_{\mathbf{x}' \in RVE} \frac{1}{N}\mathbb{C}^I. \tag{44}$$

where N is the total number of inclusions. At this stage, for the inclusions the strain field satisfies deformation compatibility while the stress field does't satisfy the stress equilibrium.

4.2.1 Stress and strain increment of matrix material points

The polarized stress increment field $\Delta\boldsymbol{\sigma}^I - \Delta\boldsymbol{\sigma}^M$ can cause a strain increment field in the matrix. Because the matrix material is homogeneous and suffering a periodic boundary condition, this strain increment can be calculated efficiently with help of Green's function and discrete Fourier transformation. With the help of the delta function

$$\delta(\mathbf{x} - \mathbf{x}') = \begin{cases} 1 & \mathbf{x} = \mathbf{x}' \\ 0 & \mathbf{x} \neq \mathbf{x}' \end{cases} \tag{45}$$

a unit force $\delta_{mi}(\mathbf{x} \quad \mathbf{x}')$ in m plane along i direction applying at $\mathbf{x}'$ will cause a displacement field $G_{km}(\mathbf{x} - \mathbf{x}')$ at $\mathbf{x}$ satisfying the stress equilibrium

$$\bar{\mathbb{C}}_{ijkl} G_{km,lj} + \delta_{mi} = 0. \tag{46}$$

In order to solve equation 46 we have to transfer it into the frequency space

$$-\bar{\mathbb{C}}_{ijkl} \hat{G}_{km} \xi_l \xi_j + \delta_{mi} = 0 \tag{47}$$

where ξ represents the frequency. Through defining a second order tensor $A_{ik} = \bar{\mathbb{C}}_{ijkl}\xi_l\xi_j$ we find the displacement $\hat{G}_{km}$ in the frequency space

$$A_{ik}\hat{G}_{km} = \delta_{im} \quad or \quad \hat{\mathbf{G}} = \mathbf{A}^{-1}. \tag{48}$$

When one transfers back $\hat{\mathbf{G}}$ from the frequency space to the real physical space one can get the solution $\mathbf{G}$ of equation 46. However, this is not necessary because we will solve the displacement field caused by the polarised stress in the frequency space.

Based on the solution of equation 46, the displacement $\Delta \mathbf{u}^M$ with respect to the polarised stress field $\Delta\boldsymbol{\sigma}^I - \Delta\boldsymbol{\sigma}^M$ can be calculated with the help of Gauss's Theorem and the periodic arrangement of the RVE

$$\Delta u_k^M = \sum_{\mathbf{x}' \in RVE} G_{km,n} \left(\Delta\sigma_{mn}^I - \Delta\sigma_{mn}^M \right) \Omega \tag{49}$$

where Ω is the volume of the regular element inside the RVE. Furthermore we can calculate the gradient of the displacement

$$\Delta u_{k,o}^M = \sum_{\mathbf{x}' \in RVE} G_{km,no} \left(\Delta\sigma_{mn}^I - \Delta\sigma_{mn}^M \right) \Omega \tag{50}$$

and transfer equation 50 into the frequency space, with the help of the convolution theorem:

$$\Delta \hat{u}_{k,o}^M = - \sum_{\mathbf{x}' \in RVE} \hat{G}_{km} \xi_n \xi_o \left(\Delta\hat{\sigma}_{mn}^I - \Delta\hat{\sigma}_{mn}^M \right) \Omega. \tag{51}$$

Through taking equation 48 into equation 51 we can calculate the displacement gradient in the frequency space

$$\begin{aligned} \Delta \hat{u}_{k,o}^M &= - \sum_{\mathbf{x}' \in RVE} A_{km}^{-1} \xi_n \xi_o \left(\Delta\hat{\sigma}_{mn}^I - \Delta\hat{\sigma}_{mn}^M \right) \Omega \\ &= - \sum_{\mathbf{x}' \in RVE} \hat{\mathbb{S}}_{komn} \left(\Delta\hat{\sigma}_{mn}^I - \Delta\hat{\sigma}_{mn}^M \right) \Omega. \end{aligned} \tag{52}$$

where $\hat{\mathbb{S}}_{komn}$ is the compliance tensor in the frequency space for the material point at the real space $\mathbf{x}'$. Finally, after one transfers equation 52 from frequency space back to the real physical space one gets the stress and strain increments for the matrix material points

$$\Delta\epsilon_{ij}^M = \frac{1}{2}(\Delta u_{i,j}^M + \Delta u_{j,i}^M) \tag{53}$$

$$\Delta\sigma_{ij}^M = \bar{\mathbb{C}}_{ijkl}^{-1} \Delta\epsilon_{kl}^M. \tag{54}$$

4.2.2 Stress and strain increment of inclusions

For each material point at $\mathbf{x}$ when there is a strain misfit between matrix and inclusion $\Delta\epsilon^I \neq \Delta\epsilon^M$ there will be a internal misfit stress $\bar{\mathbb{C}} \left(\Delta\epsilon^I - \Delta\epsilon^M \right)$. The total strain energy increment amounts to

$$\Delta E = \sum_{\mathbf{x}' \in RVE} \left[\left(\boldsymbol{\sigma}^I + \Delta\boldsymbol{\sigma}^I \right) \cdot \Delta\epsilon^I + \bar{\mathbb{C}} \left(\Delta\epsilon^I - \Delta\epsilon^M \right) \cdot \left(\Delta\epsilon^I - \Delta\epsilon^M \right) \Omega \right]. \tag{55}$$

Now the procedure of finding stress equilibrium can be replaced with the procedure of achieving the total strain energy increment minimisation

$$\nabla \cdot \boldsymbol{\sigma}^I = \mathbf{0} \iff \frac{\partial \Delta E}{\partial \Delta\epsilon^I} = 0. \tag{56}$$

Because each inclusion only has interaction with the matrix, for a fixed matrix property, material points at $\mathbf{x}$ and at $\mathbf{x}'$ are independent if $\mathbf{x} \neq \mathbf{x}'$. Under this condition, for each material point we can calculate the local strain increment which can keep the total strain energy increment minimisation

$$\Delta\epsilon^I = \left(\mathbb{C}^I + \bar{\mathbb{C}}\right)^{-1}\left(\bar{\mathbb{C}}\Delta\epsilon^M - \boldsymbol{\sigma}^I - \Delta\boldsymbol{\sigma}^I\right) \tag{57}$$

and with help of equation 57, the stress increment $\Delta\boldsymbol{\sigma}$ of inclusion at $\mathbf{x}'$ can be easily recalculated by the constitutive law.

4.2.3 Deformation compatibility

In Lebensohn (2001) the deformation compatibility problem and the energy minimisation problems are joined together through adopting the Lagrange multiplier method. In this work, after the stress and strain calculation for inclusions we simply set the matrix material point deformation increment as

$$\Delta\epsilon^M = \Delta\epsilon^I. \tag{58}$$

The final stress and strain solution for inclusions satisfying stress equilibrium and strain compatibility will not be achieved until

$$\sum_{\mathbf{x}' \in RVE} \frac{1}{N}\left\|\Delta\epsilon_k^I - \Delta\epsilon_{k+1}^I\right\| \leq C_{rsd} \tag{59}$$

where k and $k+1$ are iteration numbers and C_{rsd} the critical residual.

5. Instructive examples

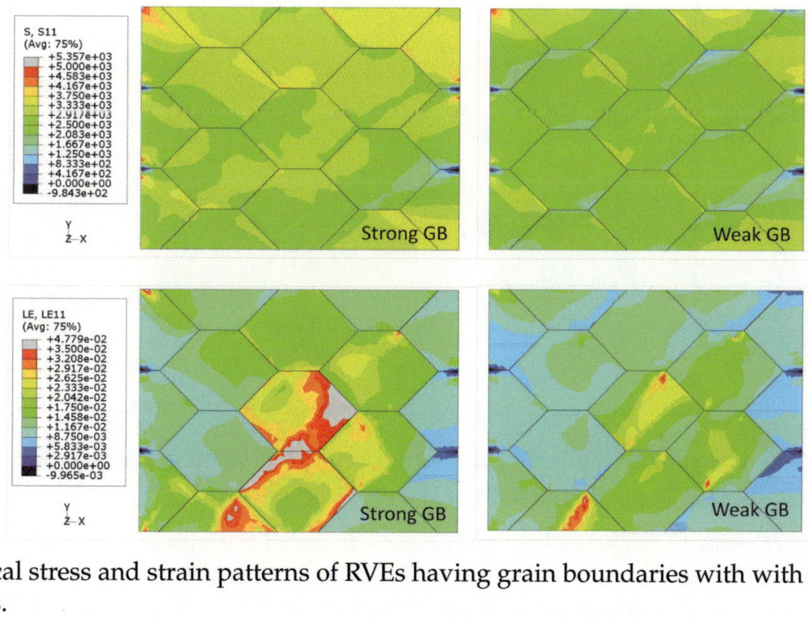

Fig. 6. Local stress and strain patterns of RVEs having grain boundaries with with different properties.

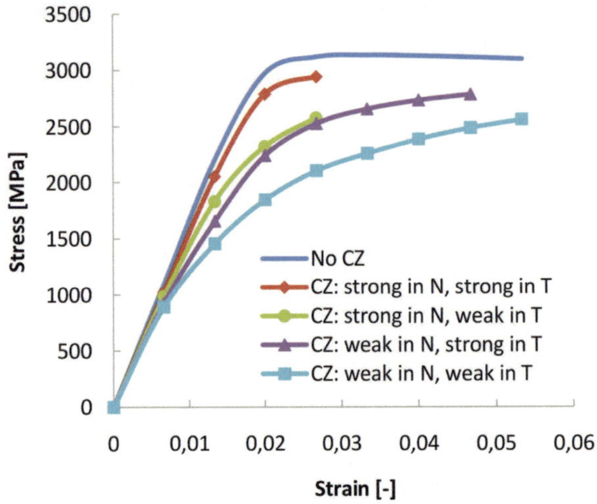

Fig. 7. Global stress-strain curves of RVEs having grain boundaries with with different properties.

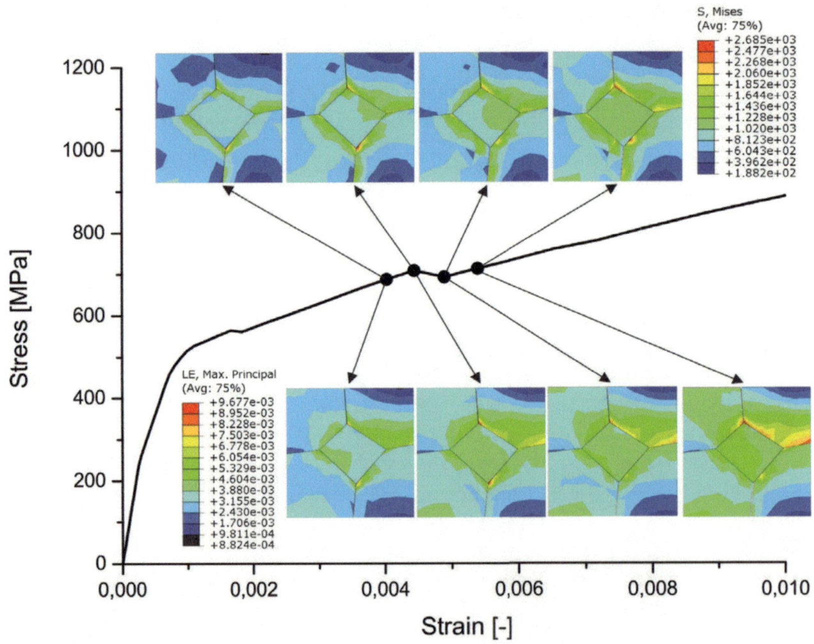

Fig. 8. Global stress-strain curve and local stress and strain distribution.

5.1 Polycrystal deformation modelling with bulk material slip and grain boundary material glide with the crystal plasticity method

A qusi-2D RVE with 17 hexagonal shape grains with 80nm $\times$ 60nm $\times$ 2nm volume has been generated. Keeping $\phi = 0$ and $\varphi_2 = 0$ we have assigned initial crystal orientations with an 5° increment for Euler angle φ_1 from grain 0° to 80° randomly. For the studied aluminum polycrystal slip systems with $a/2[1,1,0]$ and $a/2[-1,1,0]$ Burgers vectors will be activated under tensile loads along the horizontal direction (Shaban et al., 2010).

Figure 6 shows the stress and strain distributions at about 2% tensile strain with a loading speed of 10 nm/s. These results are calculated with combinations of a normal interface strength of 1500 MPa and two tangential interface strengths; an strong one of 1500 MPa and an weak one of 300 MPa. One can easily see that the stronger grain boundary causes higher stress concentrations and strain heterogeneities inside the aggregate compared with the weaker grain boundary. For the weaker grain boundary, the strain localisation instead starts from the triple junction and tends to expand into the bulk material roughly along the maximum shear stress direction. Although this phenomenon is also observed for the stronger grain boundaries, the most obvious strain localisation is in some grains with a larger Schmid factor along the grain boundary direction and even extending to the grain center in some extreme cases. In both cases, cracks have initiated in the upper and lower boundaries and attempted to propagate along the vertical grain boundaries under tensile loading along the horizontal direction.

Figure 7 shows the global stress-strain curves with respect to 5 different grain boundary strength conditions. From this plot, one can see that the combination of grain boundary opening and sliding can relax almost one third of the average stress level. Because the weak-normal-strong-tangential cohesive zone model and strong-normal-weak-tangential cohesive zone model produce almost the same global stress strain curve, it seems that the normal and tangential cohesive strengths have similar influences on the material load carrying capacity.

With respect to an small qusi-2D RVE with 5 grains in a $1\mu m \times 1\mu m$ domain, the details of the global stress strain curve have been studied as shown in Figure 8. The RVE shows some unstable mechanical behaviors especially at about 0.5% tensile strain. Generally the material load carrying capacity loss infers the damage initialisation. From our calculations, one can see clearly that the kinks of the stress strain curve are mainly stemming from grain boundary opening and sliding near grain boundary triple junctions which can relax the locally accumulated stress efficiently.

The current study with respective to nano-metere grain size polycrystals implies that grain boundary mediated deformation processes decisively change the global stress-strain response of the studied material. Since the grain boundary cohesive behavior is independent of the grain size, whereas the resistance for dislocation slip inside the bulk material points becomes smaller as grain size increase, we expect that the influence of grain boundary processes will gradually vanish for coarse-grained material.

5.2 TRIP steel deformation modelling with the crystal plasticity method

An RVE including 12 ferritic grains and one austenitic grain as shown in Figure 9 has been generated to investigate the TRIP behavior under different loading conditions. As shown in Figure 10, tensile and compression loadings induce different total martensitic volume fractions. This numerical result is consistent with experimental observations. During the martensite phase transformation, about 22% shear strain inside the habit plane and about 2% dilatation strain along the normal of the habit plane are needed to transfer from the FCC

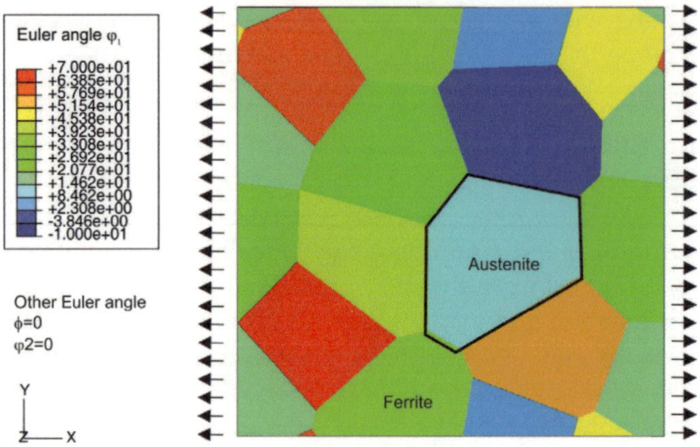

Fig. 9. Initial orientations of ferritic and austenitic grains. The only austenitic grain has been highlighted and has a volume fraction of about 10%. For the compression calculation, the loading direction has been inversed.

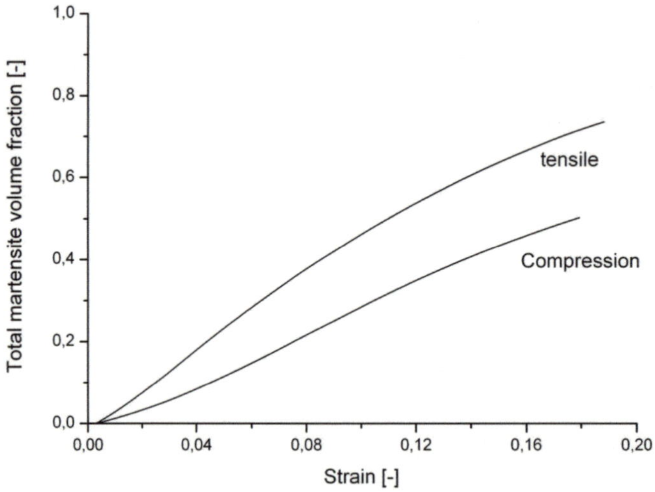

Fig. 10. Total martensitic volume fractions under tensile and compression loading cases.

lattice to the BCT lattice. Modelling results support that the normal part of the resolved stress calculated by equation 37 strongly influences the transformation volume fraction evolution. As stated in literature (Kouznetsova & Geers, 2008; Stringfellow et al., 1992), this is the well known hydrostatic stress dependence of the martensite transformation. The current study shows that there are four dominating transformation systems under tensile and compression loading conditions as shown in Figure 11. Careful analysis of the magnitude of dilatation and shear resolved stresses under tensile and compression loads shows that the shear part is important to determine the activated transformation systems.

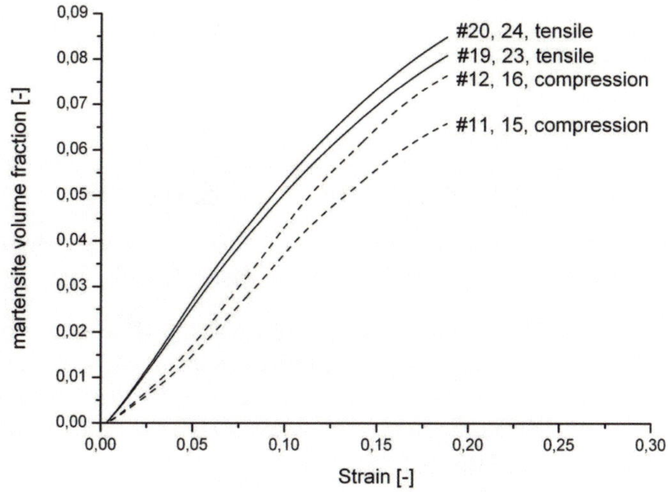

Fig. 11. Martensitic volume fractions of specific transformation systems under tensile and compression loads.

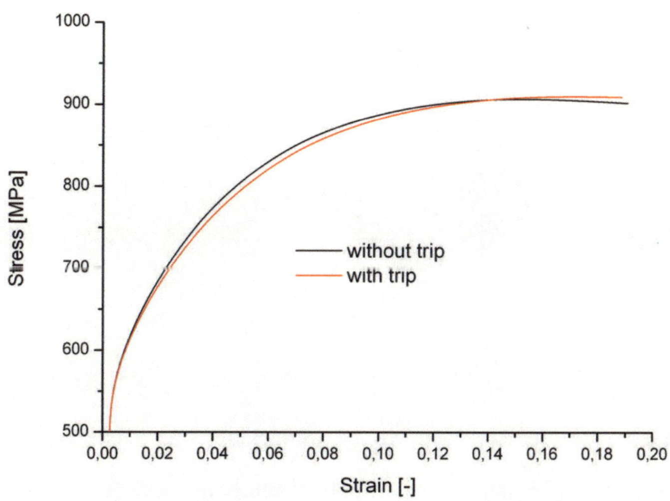

Fig. 12. Stress-strain curve comparison between deformations with and without martensite transformation.

With the help of the 13 grain RVE we have investigated into why martensite phase transformation can provide high ductility and high strength at the same time. As shown in Figure 12, the global stress-strain curves of the simulation with and without martensite transformation have a intersection point at about 14% tensile strain. Before the intersection point, the eigen strain of the phase transformation serves as a competing partner of dislocation

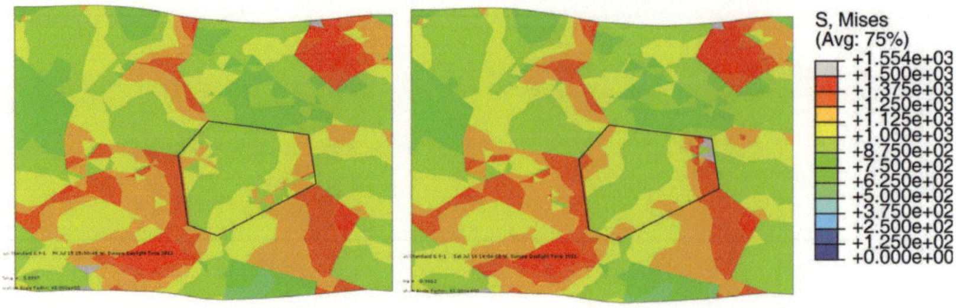

Fig. 13. Local stress distribution comparison between deformations with (right) and without (left) martensite transformation. The austenitic grain has been highlighted.

slip to reduce the external load potential, and as a direct result TRIP can increase the material ductility as shown in Figure 12. Because the phase transformation will exhaust the dislocation slip volume fraction and the martensite can only deform elastically, after the intersection point the hardening side of the TRIP mechanism will overcome the softening side and one can observe there is a enhanced tensile strength.

Figure 13 shows the local stress distribution comparison between simulations with and without phase transformation. As expected there are higher internal stresses inside the austenite grain when phase transformation exists. Through modelling the internal stress accumulation the current model system can be used to investigate the material damage and failure phenomena.

Fig. 14. Local strain distributions of an 189 grain RVE.

5.3 Polycrystal with twin lamella deformation modelling with the crystal plasticity method

As a mesh free method, the fast Fourier transformation approach can be used to model deformation of RVEs with very complicated microstructures. Several RVEs occupying a $1\mu m \times 1\mu m \times 1\mu m$ space with nano-metere sized twin lamellas inside nano-metere sized grains have been generated and discretised to $64 \times 64 \times 64$ regular grids. The initial crystal orientations have been assigned randomly. Based on equations 5 and 6, twin lamellas with different thickness have been generated. Figure 14 shows an RVE with 189 grains containing a 31 nm thickness lamella and the von-Mises equivlent strain distribution under tensile load and periodic boundary conditions. Here, the material parameters of pure aluminum have been used during the simulation.

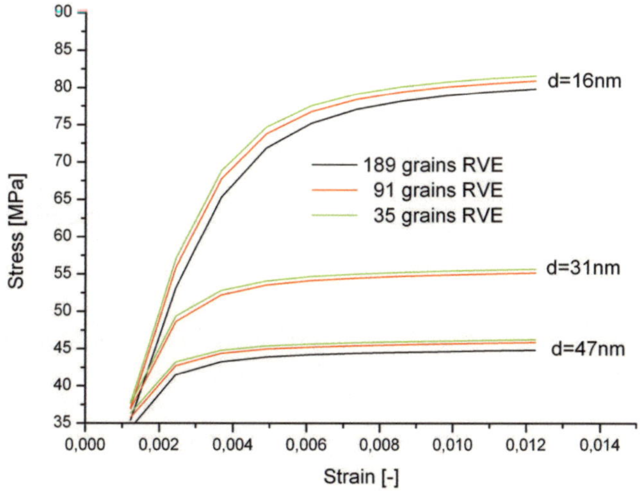

Fig. 15. Stress-strain curve comparison among RVEs with different grain size and lamella thickness.

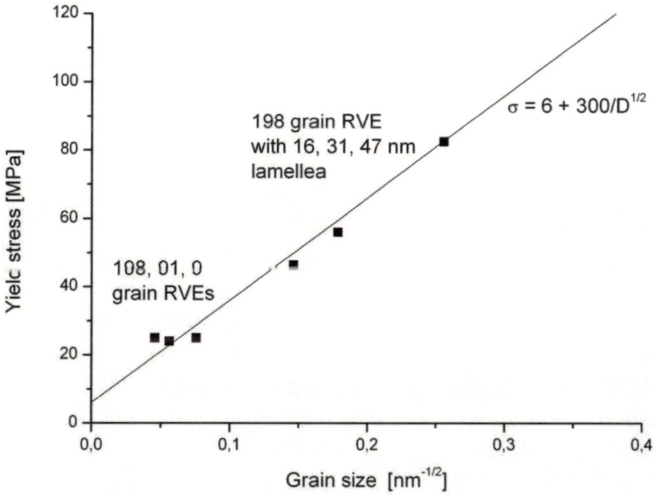

Fig. 16. Yield stress comparison among RVEs with different grain size and lamella thickness.

With respect to four grain numbers (189, 91, 35, 9) and three lamella thicknesses (16 nm, 31 nm, 47 nm), a total of 12 global stress-strain curves have been simulated to investigate the size effect on material mechanical behaviors. Figure 15 shows several stress-strain curves along the loading direction. From these results one can see easily *the smaller the stronger* rule often observed in experiments. Through determining the yield stress for different RVEs, the parameters σ_0 and k_y of the Hall-Petch relation

$$\sigma = \sigma_0 + \frac{k_y}{\sqrt{D}} \tag{60}$$

for pure aluminum have been investigated carefully. From the numerical results shown in Figure 16 we found that equation 60 with parameters σ_0=6MPa and k_y=300MPa$\sqrt{nm}$ fits the simulation data very well. Indeed, these two values almost fall nicely in the experimental measurement ranges σ_0=6 $\pm$ 2 MPa and k_y=400 $\pm$ 80MPa $\sqrt{nm}$ in Bonetti et al. (1992).

6. Summary

In this work it has been demonstrated how information from several length scales can be integrated into representative volume element (RVE) models for the mechanical behaviour of heterogeneous materials, consisting of several grains and different phases. In particular, the relevance of phenomena on different scales, like atomic bonds that determine the mechanical properties of grain boundaries, or the interaction of dislocations with grain boundaries should be investigated carefully in future studies. The mechanisms occurring at such atomistic and microstructural scales need to be modelled in a suited way such that they can be taken into account in continuum simulations of RVE's. Once an RVE for a given microstructure is constructed and the critical deformation and damage mechanisms are included into the constitutive relations, this RVE can be applied to calculate stress-strain curves and other mechanical data. The advantage of this approach is that by conducting parametric studies the influence of several microstructural features, like for example grain size or strength of grain boundaries, on the macroscopic mechanical response of a material can be predicted.

7. References

ABAQUS (2009). *ABAQUS Version6.91*, Dassault Systemes.

Bhadeshia, K. (2002). Trip-assisted steels?, *ISIJ International* 42: 1059.

Bhattacharya, K. (1993). Comparison of the geometrically nonlinear and linear theories of martensitic-transformation, *Continuum Mech. Therm.* 5: 205.

Bonetti, E., Pasquini, L. & Sampaolesi, E. (1992). The influence of grain size on the mechanical properties of nanocrystalline aluminium, *Nanostructured Materials* 9: 611.

Dai, H. & Parks, D. (1997). Geometrically-necessary dislocation density and scale-dependent crystal plasticity, *Khan, A., (Ed.),Proceedings of Sixth International Symposium on Plasticity, Gordon and Breach* .

Devincre, B., Hoc, T. & Kubin, L. (2008). Dislocation mean free paths and strain hardening of crystals, *Science* 320: 1745.

Gottstein, G. (2004). *Physical Foundations of Materials Science*, Springer Verlag, Berlin-Heidelberg, Germany.

Hane, K. & Shield, T. (1998). Symmetry and microstructure in martensites, *Philosophical Magazine A* 78.

Hane, K. & Shield, T. (1999). Microstructure in the cubic to monoclinic transition in titanium-nickel shape memory alloys, *Acta Mater.* 47.

Hirth, J. & Lothe, J. (1992). *Theory of dislocations*, Krieger Pub Co.

Janisch, R., Ahmed, N. & Hartmaier, A. (2010). Ab initio tensile tests of Al bulk crystals and grain boundaries: universality of mechanical behavior, *Physical Review B* 81: 184108–1–6.

Kalidindi, S., Bronkhort, C. & Anand, L. (1992). Crystallographic texture evolution in bulk deformation processing of fcc metals, *J. Mech. Phys. Solids* 40.

Kocks, U., Argon, A. & Ashby, M. (1975). Thermodynamics and kinetics of slip, *Chalmers, B., Christian, J.W., Massalski, T.B. (Eds.), Progress in Materials Science* 19: 1–289.

Kouznetsova, V. & Geers, M. (2008). A multi-scale model of martensitic transformation plasticity, *Mechanics of Materials* 40: 641.

Lebensohn, R. A. (2001). N-site modeling of a 3d viscoplastic polycrystal using fast fourier transform, *Acta Materialia* 49: 2723.

Lee, E. (1969). Elastic-plastic deformation at finite strains, *J Appl. Mech.* 36: 1–6.

Lu, K., Lu, L. & Suresh, S. (2009). Strengthening materials by engineering coherent internal boundaries at the nanoscale, *Science* 324: 349.

Lu, L., Shen, Y., Chen, X., Qian, L. & Lu, K. (2004). Ultrahigh strength and high electrical conductivity in copper, *Science* 304: 422.

Ma, A. (2006). *Modeling the constitutive behavior of polycrystalline metals based on dislocation mechanisms*, Phd thesis, RWTH Aachen.

Ma, A. & Roters, F. (2004). A constitutive model for fcc single crystals based on dislocation densities and its application to uniaxial compression of aluminium single crystals, *Acta Materialia* 52: 3603–3612.

Madec, R., Devincre, B., Kubin, L., Hoc, T. & Rodney, D. (2008). The role of collinear interaction in dislocation-induced hardening, *Science* 301: 1879.

Michel, J., Moulinec, H. & Suquet, P. (2000). A computational method based on augmented lagrangians and fast fourier transforms for composites with high contrast, *Comput. Model. Eng. Sci.* 1: 79.

Michel, J., Moulinec, H. & Suquet, P. (2001). A computational scheme for linear and non-linear composites with arbitrary phase contrast, *Int. J. Numer. Methods Eng.* 52: 139.

Nemat-Nasser, S., Luqun, N. & Okinaka, T. (1998). A constitutive model for fcc crystals with application to polycrystalline ofhc copper, *Mech. Mater.* 30: 325–341.

Nye, J. (1953). Some geometrical relations in dislocated crystals, *Acta Metall.* 1: 153–162.

Olson, G. & Cohen, M. (1972). A mechanism for strain-induced nucleation of martensitic transformations, *J. Less-Common Metals* 28.

Peirce, D., Asaro, R. & Needleman, A. (1982). An analysis of non-uniform and localized deformation in ductile single crystals, *Acta Metall.* 30: 1087–1119.

Roters, F. (1999). *Realisierung eines Mehrebenenkonzeptes in der Plastizitätsmodellierung*, Phd thesis, RWTH Aachen.

Schröder, J., Balzani, D. & Brands, D. (2010). Approximation of random microstructures by periodic statistically similar representative volume elements based on lineal-path functions, *Archive of Applied Mechanics* 81: 975.

Shaban, A., Ma, A. & Hartmaier, A. (2010). Polycrystalline material deformation modeling with grain boundary sliding and damage accumulation, *Proceedings of 18th European Conference on Fracture (ECF18)* .

Stringfellow, R., Parks, D. & Olson, G. (1992). A constitutive model for transformation plasticity accompanying strain-induced martensitic transformation in metastable austenitic steels, *Acta Metall. Mater.* 40: 1703.

Tjahjanto, D. D. (2008). *Micromechanical modeling and simulations of transformation-induced plasticity in multiphase carbon steels*, Phd thesis, Delft University of Technology.

Wechsler, M., Lieberman, T. & Read, T. (1953). On the theory of the formation of martensite, *Trans. AIME J. Metals* 197: 1094.

Wei, Y. & Anand, L. (2004). Grain-boundary sliding and separation in polycrystalline metals: application to nanocrystalline fcc metals, *Journal of the Mechanics and Physics of Solids* 52: 2587.

2

Strength of a Polycrystalline Material

P.V. Galptshyan
Institute of Mechanics, National Academy of Sciences of the Republic of Armenia, Erevan Republic of Armenia

1. Introduction

There are numerous polycrystalline materials, including polycrystals whose crystals have a cubic symmetry. Polycrystals with cubic symmetry comprise minerals and metals such as cubic pyrites (FeS2), fluorite (CaF2), rock salt (NaCl), sylvite (KCl), iron (Fe), aluminum (Al), copper (Cu), and tungsten (W) (Love, 1927; Vainstein et al., 1981).

It is assumed that many materials can be treated as a homogeneous and isotropic medium independently of the specific characteristics of their microstructure. It is clear that, in fact, this is impossible already because of the molecular structure of materials. For example, materials with polycrystalline structure, which consist of numerous chaotically located small crystals of different size and different orientation, cannot actually be homogeneous and isotropic. Each separate crystal of the metal is anisotropic. But if the volume contains very many chaotically located crystals, then the material as a whole can be treated as an isotropic material. Just in a similar way, if the geometric dimensions of a body are large compared with the dimensions of a single crystal, then, with a high degree of accuracy, one can assume that the material is homogeneous (Feodos'ev, 1979; Timoshenko & Goodyear, 1951).

On the other hand, if the problem is considered in more detail, then the anisotropy both of the material and of separate crystals must be taken into account. For a body under the action of external forces, it is impossible to determine the stress-strain state theoretically with its polycrystalline structure taken into account.

Assume that a body consists of crystals of the same material. Moreover, in general, the principal directions of elasticity of neighboring crystals do not coincide and are oriented arbitrarily. The following question arises: Can stress concentration exist near a corner point of the interface between neighboring crystals and near and edge of the interface?

To answer this question, it is convenient to replace the problem under study by several simplified problems each of which can reflect separate situations in which several neighboring crystals may occur.

A similar problem for two orthotropic crystals having the shape of wedges rigidly connected along their jointing plane was considered in (Belubekyan, 2000). They have a common vertex, and their external faces are free. Both of the wedges consist of the same material. The wedges have common principal direction of elasticity of the same name, and the other elastic-equivalent principal directions form a nonzero angle. We consider longitudinal shear (out-of-plane strain) along the common principal direction.

In (Belubekyan, 2000), it is shown that if the joined wedges consist of the same orthotropic material but have different orientations of the principal directions of elasticity with respect to their interface, then the compound wedge behaves as a homogeneous wedge.

The behavior of the stress field near the corner point of the contour of the transverse cross-section of the compound body formed by two prismatic bodies with different characteristics which are welded along their lateral surfaces was studied in the case of plane strain in (Chobanyan, 1987). It was assumed there that the compound parts of the body are homogeneous and isotropic and the corner point of the contour of the prism transverse cross-section lies at the edge of the contact surface of the two bodies.

In (Chobanyan, 1987; Chobanyan & Gevorkyan 1971), the character of the stress distribution near the corner point of the contact surface is also studied for two prismatic bodies welded along part of their lateral surfaces. The plane strain of the compound prism is considered.

There are numerous papers dealing with the mechanics of contact interaction between strained rigid bodies. The contact problems of elasticity are considered in the monographs (Alexandrov & Romalis, 1986; Alexandrov & Pozharskii 1998). In (Alexandrov & Romalis, 1986), exact or approximate analytic solutions are obtained in the form convenient to be used directly to verify the contact strength and rigidity of machinery elements. The monograph (Alexandrov & Pozharskii 1998) presents numericalanalytical methods and the results of solving many nonclassical spatial problems of mechanics of contact interaction between elastic bodies. Isotropic bodies of semibounded dimensions (including the wedge and the cone) and the bodies of bounded dimensions were considered. The monograph presents a vast material developed in numerous publications. There are also many studies in this field, which were published in recent years (Ulitko & Kochalovskaya, 1995; Pozharskii & Chebakov, 1998; Alexandrov & Pozharskii, 1998, 2004; Alexandrov et al., 2000; Osrtrik & Ulitko, 2000; Alexandrov & Klindukhov, 2000, 2005; Pozharskii, 2000, 2004; Aleksandrov, 2002, 2006; Alexandrov & Kalyakin, 2005).

In the present paper, we study the problem of existence of stress concentrations near the corner point of the interface between two joined crystals with cubic symmetry made of the same material.

2. Statement of the problem

We assume that there are two crystals with rectilinear anisotropy and cubic symmetry, which are rigidly connected along their contact surface (Fig. 1). The crystal contact surface forms a dihedral angle with linear angle α whose trace is shown in the plane of the drawing. The contact surface edge passes through point O. The z-axis of the cylindrical coordinate system (r,φ,z) coincides with the edge of the dihedral angle. The coordinate surfaces and $\varphi=0$ and $\varphi=\alpha$ $(\varphi=\alpha-2\pi)$ coincide with the faces of the dihedral angle. Thus, the first crystal (1) occupies the domain $\varphi\in[0;\alpha]$ and the second crystal (2) occupies the domain $\varphi\in[\alpha-2\pi;\,0]$. In this case $0<\alpha<2\pi$ and $0<r<\infty$.

For simplicity, we assume that the crystals have a single common principal direction of elasticity coinciding with the z-axis. The other two principal directions x_1 and y_1 of the first crystal make some nonzero angles with the principal directions x_2 and y_2 of the second crystal. By θ_1 we denote the angle between x_1 and the polar axis $\varphi = 0$, and by θ_2, the angle between x_2 and the axis $\varphi = 0$. In this case, $\theta_1, \theta_2 \subset [\alpha - 2\pi, \alpha]$. If $\theta_1 = \theta_2 = 0$, then we have a homogeneous medium, i.e., a monocrystal with cubic symmetry, one of whose principal directions $x_1 = x_2 = x$ coincides with the polar axis $\varphi = 0$. In this case, the equations of generalized Hooke's law written in the principal axes of elasticity x, y, z have the form

$$\begin{aligned} \varepsilon_x &= a_{11}\sigma_x + a_{12}\left(\sigma_y + \sigma_z\right), & \gamma_{yz} &= a_{44}\tau_{yz}, \\ \varepsilon_y &= a_{11}\sigma_y + a_{12}\left(\sigma_z + \sigma_x\right), & \gamma_{zx} &= a_{44}\tau_{zx}, \\ \varepsilon_z &= a_{11}\sigma_z + a_{12}\left(\sigma_x + \sigma_y\right), & \gamma_{xy} &= a_{44}\tau_{xy}, \end{aligned} \quad (1)$$

where $\varepsilon_x, \varepsilon_y, ..., \gamma_{xy}$ are the strain components, $\sigma_x, \sigma_y, ..., \tau_{xy}$ are the stress components, and a_{11}, a_{12}, a_{44} are the strain coefficients.

Equations (1) can be obtained from the equations of generalized Hooke's law for an orthotropic body written in the principal axes of elasticity x, y, z, using the method described in (Lekhnitskii, 1981).

Rotating the coordinate system (x, y, z) about the common axis $z = z^{/}$ by the angle $\varphi = 90^\circ$, we obtain a symmetric coordinate system (x', y', z'). Since the directions of the axes x, y, z and $x^{/}, y^{/}, z^{/}$ of the same name are equivalent with respect to their elastic properties, the equations of generalized

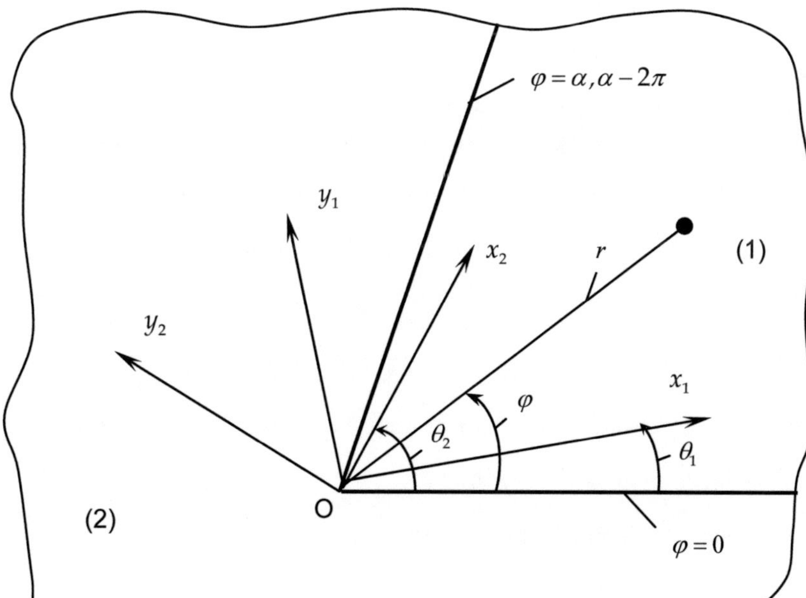

Fig. 1.

Hooke's law for these coordinate systems have the same form. In this case, the values of the strain coefficients are the same in both systems: $a'_{11} = a_{11}$, $a'_{12} = a_{12}$, $a'_{13} = a_{13}$, ... ,$a'_{66} = a_{66}$.

Using the formulas of transformation of strain coefficients under the rotation of the coordinate system about the axis $z = z'$ (Lekhnitskii, 1981), we obtain their new values expressed in terms of the old values (before the rotation of the coordinate system (x, y, z)).

Comparing the strain coefficients in the same coordinate system (x', y', z'), we obtain, $a_{11} = a_{12}$, $a_{44} = a_{55}$, $a_{13} = a_{23}$, $a_{16} = a_{45} = a_{26} = a_{36} = 0$.

Successively rotating the coordinate system (x, y, z) about the axes x and y by the angle 90° and repeating the same procedure, we finally obtain (1).

The transformation formulas for the strain coefficients under the rotation of the coordinate system about the x-and y-axes can also be obtained from the transformation formulas for the strain coefficients under the rotation of the coordinate system about the z-axis in the case of anisotropy of general form.

For example, to obtain the transformation formulas under the rotation of the coordinate system about the x-axis, it is necessary to rename the principal directions of elasticity as follows: the x-axis becomes the z-axis, the y-axis becomes the x-axis, and the z-axis becomes the y-axis. In this case, in the equations of generalized Hooke's law referred to the coordinate system (x, y, z), a_{22} plays the role of a_{11}, a_{23} plays the role of a_{12}, and a_{24} plays the role of a_{16}. In a similar way, in the equations of generalized Hooke's law referred to the coordinate system (x', y', z'), a'_{22} plays the role of a'_{11}, a'_{23} plays the role of a'_{12}, and a'_{24} plays the role of a'_{16}. This implies that, in the case of an orthotropic body, $a_{24} = 0$ under rotation of the coordinate system about the x-axis, but, in contrast to the case of rotation of the coordinate system about the z-axis, a'_{24} is generally nonzero.

In the case $\theta_1 \neq \theta_2$, the equations of generalized Hooke's law in the cylindrical coordinate system (r, φ, z) have the form

$$
\begin{aligned}
&\varepsilon_r^{(i)} = a_{11}\sigma_r^{(i)} + a_{12}(\sigma_\varphi^{(i)} + \sigma_z^{(i)}) - a[\,(\sigma_r^{(i)} - \sigma_\varphi^{(i)})\sin^2 2\alpha_i + \tau_{r\varphi}^{(i)}\sin 4\alpha_i\,],\\
&\varepsilon_\varphi^{(i)} = a_{11}\sigma_\varphi^{(i)} + a_{12}(\sigma_z^{(i)} + \sigma_r^{(i)}) + a[\,(\sigma_r^{(i)} - \sigma_\varphi^{(i)})\sin^2 2\alpha_i + \tau_{r\varphi}^{(i)}\sin 4\alpha_i\,],\\
&\varepsilon_z^{(i)} = a_{11}\sigma_z^{(i)} + a_{12}(\sigma_r^{(i)} + \sigma_\varphi^{(i)}),\\
&\gamma_{\varphi z}^{(i)} = a_{44}\tau_{\varphi z}^{(i)} = 2(a_{11} - a_{12})\,\tau_{\varphi z}^{(i)} - 4a\tau_{\varphi z}^{(i)},\\
&\gamma_{zr}^{(i)} = a_{44}\tau_{zr}^{(i)} = 2(a_{11} - a_{12})\,\tau_{rz}^{(i)} - 4a\tau_{zr}^{(i)},\\
&\gamma_{r\varphi}^{(i)} = 2(a_{11} - a_{12})\,\tau_{r\varphi}^{(i)} - a[\,(\sigma_r^{(i)} - \sigma_\varphi^{(i)})\sin 4\alpha_i + 4\tau_{r\varphi}^{(i)}\cos^2 2\alpha_i\,],\\
&4a = 2(a_{11} - a_{12}) - a_{44}, \quad \alpha_i = \varphi - \theta_i,
\end{aligned}
\tag{2}
$$

where the above form of anisotropy is used. From now on, the first crystal is denoted by the index $i=1$, and the second, by $i=2$.

In the case of cubic symmetry of the material, we have the following dependencies between the moduli of elasticity A_{11}, A_{12}, A_{44} and the strain coefficients a_{11}, a_{12}, a_{44}:

$$A_{11} = \frac{a_{11}+a_{12}}{(a_{11}-a_{12})(a_{11}+2a_{12})}, \qquad A_{12} = -\frac{a_{12}}{(a_{11}-a_{12})(a_{11}+2a_{12})}, \qquad A_{44} = \frac{1}{a_{44}}$$

In the isotropic medium, we have $a_{44} = 2(a_{11}-a_{12})$ and $2A_{44} = A_{11}-A_{12}$. For cubic crystals, the ratio $\eta = 2A_{44}/(A_{11}-A_{12})$ is called a parameter of elastic anisotropy in (Vainstein et al., 1981). In contrast to η, we call a the coefficient of elastic anisotropy. For $a=0$, we have an anisotropic medium in Eqs. (2).

We also note that for $\varphi=\theta_i=0$, Eqs. (2) correspond to generalized Hooke's law written for monocrystals and referred to the principal axes of elasticity.

3. Out-of-plane strain

In the case of longitudinal shear along the direction of the axis z, we have the following components of the displacement vector: $u_r^{(i)} \equiv 0$, $u_\varphi^{(i)} \equiv 0$, $u_z^{(i)} = u_z^{(i)}(r,\varphi)$.

For small strains, the strain components $\gamma_{\varphi z}^{(i)}$ and $\gamma_{rz}^{(i)}$, not identically zero, are related to $u_z^{(i)}$ by the Cauchy equations: $\gamma_{\varphi z}^{(i)} = \partial u_z^{(i)}/r\partial\varphi$, $\gamma_{rz}^{(i)} = \partial u_z^{(i)}/\partial r$. According to Hooke's law (2), this implies that

$$\sigma_r = \sigma_\varphi = \sigma_z = \tau_{r\varphi} \equiv 0,$$
$$\tau_{\varphi z}^{(i)} = \frac{1}{a_{44}^{(i)}}\frac{1}{r}\frac{\partial u_z^{(i)}}{\partial\varphi}, \qquad \tau_{rz}^{(i)} = \frac{1}{a_{44}^{(i)}}\frac{\partial u_z^{(i)}}{\partial r}. \tag{3}$$

Substituting (3) into the differential equations of equilibrium, we obtain $\Delta u_z^{(i)} = 0$, where Δ is the Laplace operator.

Since the crystals are rigidly joined, on the interface between the two crystals the displacements are continuous,

$$u_z^{(1)}(r,0) = u_z^{(2)}(r,0), \quad u_z^{(1)}(r,\alpha) = u_z^{(2)}(r,\alpha-2\pi),$$

and the contact stresses are continuous,

$$\frac{\partial u_z^{(1)}(r,0)}{\partial\varphi} = \frac{a_{44}^{(1)}}{a_{44}^{(2)}}\frac{\partial u_z^{(2)}(r,0)}{\partial\varphi}, \quad \frac{\partial u_z^{(1)}(r,\alpha)}{\partial\varphi} = \frac{a_{44}^{(1)}}{a_{44}^{(2)}}\frac{\partial u_z^{(2)}(r,\alpha-2\pi)}{\partial\varphi}.$$

Since $a_{44}^{(1)} = a_{44}^{(2)} = a_{44}$, this implies that, in the case of out-of-plane strain, the two-crystal composed of monocrystals of the same material behaves as a monocrystal corresponding to the case $\theta_1 = \theta_2$.

Thus, in the case of longitudinal shear in the direction of the z -axis , there is no stress concentration at the corner point of the interface between the two joined crystals regardless of the orientation of the principal directions x_1 and x_2 .

4. Plane strain

In this case, we have

$$u_r^{(i)} = u_r^{(i)}(r,\varphi), \quad u_\varphi^{(i)} = u_\varphi^{(i)}(r,\varphi), \quad u_z^{(i)} \equiv 0.$$

Hence the following strain components are nonzero:

$$\varepsilon_r^{(i)} = \frac{\partial u_r^{(i)}}{\partial r}, \quad \varepsilon_\varphi^{(i)} = \frac{1}{r}\frac{\partial u_\varphi^{(i)}}{\partial \varphi} + \frac{u_r^{(i)}}{r}, \quad r\gamma_{r\varphi}^{(i)} = \frac{\partial u_r^{(i)}}{\partial \varphi} + r\frac{\partial u_\varphi^{(i)}}{\partial r} - u_\varphi^{(i)}. \tag{4}$$

Hooke's law (2) has the form

$$\begin{aligned}
&\varepsilon_r^{(i)} = b_1\,\sigma_r^{(i)} + b_2\,\sigma_\varphi^{(i)} - a\left[\,(\sigma_r^{(i)} - \sigma_\varphi^{(i)})\sin^2 2\alpha_i + \tau_{r\varphi}^{(i)}\sin 4\alpha_i\,\right],\\
&\varepsilon_\varphi^{(i)} = b_2\,\sigma_r^{(i)} + b_1\sigma_\varphi^{(i)} + a\left[\,(\sigma_r^{(i)} - \sigma_\varphi^{(i)})\sin^2 2\alpha_i + \tau_{r\varphi}^{(i)}\sin 4\alpha_i\,\right],\\
&\gamma_{r\varphi}^{(i)} = 2(a_{11} - a_{12})\,\tau_{r\varphi}^{(i)} - a\left[\,(\sigma_r^{(i)} - \sigma_\varphi^{(i)})\sin 4\alpha_i + 4\tau_{r\varphi}^{(i)}\cos^2 2\alpha_i\,\right],\\
&b_1 = a_{11} - \frac{a_{12}^2}{a_{11}}, \quad b_2 = a_{12} - \frac{a_{12}^2}{a_{11}}.
\end{aligned} \tag{5}$$

In the absence of mass forces, we satisfy the differential equations of equilibrium by expressing $\sigma_r^{(i)}$, $\sigma_\varphi^{(i)}$ and $\tau_{r\varphi}^{(i)}$ via the Airy stress function Φ_i :

$$\sigma_r^{(i)} = \frac{1}{r^2}\frac{\partial^2\Phi_i}{\partial\varphi^2} + \frac{1}{r}\frac{\partial\Phi_i}{\partial r}, \quad \sigma_\varphi^{(i)} = \frac{\partial^2\Phi_i}{\partial r^2}, \quad \tau_{r\varphi}^{(i)} = -\frac{\partial}{\partial r}\left(\frac{1}{r}\frac{\partial\Phi_i}{\partial\varphi}\right). \tag{6}$$

By substituting (5) into the strain consistency condition

$$\frac{\partial^2\gamma_{r\varphi}^{(i)}}{\partial r\,\partial\varphi} - r\frac{\partial^2\varepsilon_\varphi^{(i)}}{\partial r^2} - \frac{1}{r}\frac{\partial^2\varepsilon_r^{(i)}}{\partial\varphi^2} + \frac{1}{r}\frac{\partial\gamma_{r\varphi}^{(i)}}{\partial\varphi} - 2\frac{\partial\varepsilon_\varphi^{(i)}}{\partial r} + \frac{\partial\varepsilon_r^{(i)}}{\partial r} = 0$$

after several simplifying transformations, according to (6), we obtain the basic equation of the problem:

$$\Delta^2\Phi_i - \frac{a}{b_1}\left[\left(\frac{\partial^4\Phi_i}{\partial r^4} + \frac{1}{r^4}\frac{\partial^4\Phi_i}{\partial\varphi^4}\right)\sin^2 2\alpha_i - 2\left(\frac{1}{r^2}\frac{\partial^4\Phi_i}{\partial r^2\partial\varphi^2} + \frac{1}{r}\frac{\partial^3\Phi_i}{\partial r^3}\right)\right.$$
$$\times\left(3\sin^2 2\alpha_i - 2\right) + \left(\frac{2}{r^3}\frac{\partial^3\Phi_i}{\partial r\partial\varphi^2} + \frac{1}{r^2}\frac{\partial^2\Phi_i}{\partial r^2} - \frac{1}{r^3}\frac{\partial\Phi_i}{\partial r}\right)\left(15\sin^2 2\alpha_i - 8\right)$$
$$-\frac{4}{r^4}\frac{\partial^2\Phi_i}{\partial\varphi^2}\left(11\sin^2 2\alpha_i - 6\right) + 2\left(\frac{1}{r}\frac{\partial^4\Phi_i}{\partial r^3\partial\varphi} - \frac{1}{r^3}\frac{\partial^4\Phi_i}{\partial r\partial\varphi^3}\right. \qquad (7)$$
$$\left.\left. -\frac{6}{r^2}\frac{\partial^3\Phi_i}{\partial r^2\partial\varphi} + \frac{3}{r^4}\frac{\partial^3\Phi_i}{\partial\varphi^3} + \frac{14}{r^3}\frac{\partial^2\Phi_i}{\partial r\partial\varphi} - \frac{12}{r^4}\frac{\partial\Phi_i}{\partial\varphi}\right)\sin 4\alpha_i\right] = 0,$$
$$\Delta^2 = \left(\frac{\partial^2}{\partial r^2} + \frac{1}{r}\frac{\partial}{\partial r} + \frac{1}{r^2}\frac{\partial^2}{\partial\varphi^2}\right)^2.$$

The rigid connection of the crystals along their contact surface implies the continuity conditions for the displacements on this surface.

$$\frac{\partial u_r^{(1)}(r,0)}{\partial r} = \frac{\partial u_r^{(2)}(r,0)}{\partial r}, \qquad \frac{\partial^2 u_\varphi^{(1)}(r,0)}{\partial r^2} = \frac{\partial^2 u_\varphi^{(2)}(r,0)}{\partial r^2},$$
$$\frac{\partial u_r^{(1)}(r,\alpha)}{\partial r} = \frac{\partial u_r^{(2)}(r,\alpha-2\pi)}{\partial r}, \qquad \frac{\partial^2 u_\varphi^{(1)}(r,\alpha)}{\partial r^2} = \frac{\partial^2 u_\varphi^{(2)}(r,\alpha-2\pi)}{\partial r^2}. \qquad (8)$$

and the continuity conditions for the contact stresses,

$$\Phi_1(r,0) = \Phi_2(r,0), \qquad \frac{\partial\Phi_1(r,0)}{\partial\varphi} = \frac{\partial\Phi_2(r,0)}{\partial\varphi},$$
$$\Phi_1(r,\alpha) = \Phi_2(r,\alpha-2\pi), \qquad \frac{\partial\Phi_1(r,\alpha)}{\partial\varphi} = \frac{\partial\Phi_2(r,\alpha-2\pi)}{\partial\varphi}. \qquad (9)$$

If we set $a=0$ in problem (7)–(9), then we obtain a plane problem for the homogeneous isotropic body.

According to (4), (5), and (6), we have

$$\frac{\partial u_r^{(i)}}{\partial r} = b_1\left(\frac{1}{r^2}\frac{\partial^2\Phi_i}{\partial\varphi^2} + \frac{1}{r}\frac{\partial\Phi_i}{\partial r}\right) + b_2\frac{\partial^2\Phi_i}{\partial r^2}$$
$$-a\left[\left(\frac{1}{r^2}\frac{\partial^2\Phi_i}{\partial\varphi^2} + \frac{1}{r}\frac{\partial\Phi_i}{\partial r} - \frac{\partial^2\Phi_i}{\partial r^2}\right)\sin^2 2\alpha_i + \left(\frac{1}{r^2}\frac{\partial\Phi_i}{\partial\varphi} - \frac{1}{r}\frac{\partial^2\Phi_i}{\partial\varphi\partial r}\right)\sin 4\alpha_i\right], \qquad (10)$$

$$\frac{\partial u_r^{(i)}}{\partial \varphi}+r\frac{\partial u_\varphi^{(i)}}{\partial r}-u_\varphi^{(i)}=2\left(a_{11}-a_{12}\right)\left(\frac{1}{r}\frac{\partial \Phi_i}{\partial \varphi}-\frac{\partial^2\Phi_i}{\partial\varphi\partial r}\right)$$
$$+a\left[\left(r\frac{\partial^2\Phi_i}{\partial r^2}-\frac{1}{r}\frac{\partial^2\Phi_i}{\partial\varphi^2}-\frac{\partial\Phi_i}{\partial r}\right)\sin 4\alpha_i-4\cos^2 2\alpha_i\left(\frac{1}{r}\frac{\partial\Phi_i}{\partial\varphi}-\frac{\partial^2\Phi_i}{\partial\varphi\partial r}\right)\right]. \tag{11}$$

Differentiating (10) with respect to φ and (11) with respect to r and eliminating the derivative $\partial^2 u_r^{(i)}/\partial r\partial\varphi$, we obtain

$$\frac{\partial^2 u_\varphi^{(i)}}{\partial r^2}=a\left[\frac{\partial^3\Phi_i}{\partial r^3}\sin 4\alpha_i+\frac{1}{r^3}\frac{\partial^3\Phi_i}{\partial\varphi^3}\sin^2 2\alpha_i+\frac{1}{r}\frac{\partial^3\Phi_i}{\partial\varphi\partial r^2}\left(4-5\sin^2 2\alpha_i\right)\right.$$
$$-\frac{2}{r^2}\frac{\partial^3\Phi_i}{\partial\varphi^2\partial r}\sin 4\alpha_i-\frac{2}{r}\frac{\partial^2\Phi_i}{\partial r^2}\sin 4\alpha_i$$
$$+\frac{4}{r^3}\frac{\partial^2\Phi_i}{\partial\varphi^2}\sin 4\alpha_i+\frac{1}{r^2}\frac{\partial^2\Phi_i}{\partial\varphi\partial r}\left(13\sin^2 2\alpha_i-8\right)$$
$$\left.+\frac{2}{r^2}\frac{\partial\Phi_i}{\partial r}\sin 4\alpha_i+\frac{1}{r^3}\frac{\partial\Phi_i}{\partial\varphi}\left(8-12\sin^2 2\alpha_i\right)\right] \tag{12}$$
$$-b_1\frac{1}{r^3}\frac{\partial^3\Phi_i}{\partial\varphi^3}-\frac{1}{r}\frac{\partial^3\Phi_i}{\partial\varphi\partial r^2}\left(a_{11}-a_{12}+b_1\right)$$
$$+\frac{1}{r^2}\frac{\partial^2\Phi_i}{\partial\varphi\partial r}\left(a_{11}-a_{12}-b_2\right)+\frac{2}{r^3}\frac{\partial\Phi_i}{\partial\varphi}\left(a_{12}-a_{11}\right).$$

We use the expressions (10) and (12) to represent the continuity conditions (8) via the stress function Φ.

5. Solution method

For $a=0$, from (7) we derive the biharmonic equation and, solving it by separation of variables, obtain the following solution (Chobanyan, 1987; Chobanyan & Gevorkyan, 1971):

$$\Phi_i(r,\varphi)=r^{\lambda+1}F_i(\lambda;\varphi), \tag{13}$$

$$F_i(\lambda;\varphi)=\mathrm{A}_i\sin(\lambda+1)\varphi+\mathrm{B}_i\cos(\lambda+1)\varphi$$
$$+C_i\sin(\lambda-1)\varphi+D_i\cos(\lambda-1)\varphi. \tag{14}$$

where λ is a parameter and A_i, B_i, C_i and D_i –are integration constants.

For a sufficiently small in absolute value, we replace the solution of Eq. (7) by the solution of the biharmonic equation (13). By substituting (13) into (7), we obtain a fourth-order ordinary differential equation for $F_i(\lambda;\varphi)$:

$$F_i^{IV}+2\left(\lambda^2+1\right)F_i^{//}+\left(\lambda^2-1\right)^2F_i-\frac{a}{b_1}\{[F_i^{IV}+2\left(\lambda^2+1\right)F_i^{//}+\left(\lambda^2-1\right)^2F_i]\sin^2 2\alpha_i$$
$$-2(\lambda-2)\sin 4\alpha_i\,F_i^{///}+4(\lambda-1)(\lambda-2)\cos 4\alpha_i\,F_i^{//} \tag{15}$$
$$+2(\lambda-2)\left[(\lambda-2)^2-5\right]\sin 4\alpha_i\,F_i^{/}+4\left(\lambda^2-1\right)(\lambda-2)\cos 4\alpha_i\,F_i\}=0$$

whose general integral has the form (14) for $a=0$.

After the substitution of (13) into (10) and (12), we can write

$$\frac{\partial u_r^{(i)}}{\partial r}=r^{\lambda-1}\left[b_1F_i^{//}(\varphi)+(b_1+\lambda b_2)(\lambda+1)F_i(\varphi)\right]$$
$$-r^{\lambda-1}a\left[\frac{1}{2}F_i^{//}(\varphi)(1-\cos 4\alpha_i)-F_i^{/}(\varphi)\lambda\sin 4\alpha_i\right. \tag{16}$$
$$\left.-\frac{1}{2}\left(\lambda^2-1\right)F_i(\varphi)(1-\cos 4\alpha_i)\right].$$

$$\frac{\partial^2 u_\varphi^{(i)}}{\partial r^2}=r^{\lambda-2}a\left\{F_i(\varphi)\left(\lambda^2-1\right)(\lambda-2)\sin 4\alpha_i+\frac{1}{2}F_i^{/}(\varphi)\left[3\lambda^2+1+\left(5\lambda^2-8\lambda-1\right)\right.\right.$$
$$\left.\left.\times\cos 4\alpha_i\right]-2F_i^{//}(\varphi)(\lambda-1)\sin 4\alpha_i+\frac{1}{2}F_i^{///}(\varphi)(1-\cos 4\alpha_i)\right\}-r^{\lambda-2}\left\{b_1F_i^{///}(\varphi)\right. \tag{17}$$
$$\left.+F_i^{/}(\varphi)[\lambda(\lambda+1)(a_{11}-a_{12}+b_1)-(\lambda+1)(a_{11}-a_{12}-b_2)-2(a_{12}-a_{11})]\right\}.$$

According to (13), (16), and (17), the continuity conditions (8) and (9) acquire the form

$$X_{1j}=X_{2j}\qquad (j=1,2,\ldots,8),\quad X_{i1}=F_i(0),\quad X_{i2}=F_i^{/}(0), \tag{18}$$

$$X_{i3}=2b_1F_i^{//}(0)+2(b_1+\lambda b_2)(\lambda+1)F_i(0)-a\left[F_i^{//}(0)(1-\cos 4\theta_i)\right.$$
$$\left.+2F_i^{/}(0)\lambda\sin 4\theta_i-\left(\lambda^2-1\right)F_i(0)(1-\cos 4\theta_i)\right],$$

$$X_{i4}=a\left\{F_i^{///}(0)(1-\cos 4\theta_i)+4F_i^{//}(0)(\lambda-1)\sin 4\theta_i+F_i^{/}(0)\left[3\lambda^2+1\right.\right.$$
$$\left.\left.+\left(5\lambda^2-8\lambda-1\right)\cos 4\theta_i\right]-2F_i(0)\left(\lambda^2-1\right)(\lambda-2)\sin 4\theta_i\right\}-2b_1F_i^{///}(0)$$
$$+2F_i^{/}(0)\left[(\lambda+1)(a_{11}-a_{12}-b_2)-\lambda(\lambda+1)(a_{11}-a_{12}+b_1)+2(a_{12}-a_{11})\right],$$

$$X_{i5}=F_i(\beta_i),\quad X_{i6}=F_i^{/}(\beta_i),\quad \beta_1=\alpha,\quad \beta_2=\alpha-2\pi,$$

$$X_{i7}=2b_1F_i^{//}(\beta_i)+2(b_1+\lambda b_2)(\lambda+1)F_i(\beta_i)-a\left\{\left[1-\cos 4(\alpha-\theta_i)\right]F_i^{//}(\beta_i)\right.$$
$$\left.-2\lambda\sin 4(\alpha-\theta_i)F_i^{/}(\beta_i)-\left(\lambda^2-1\right)\left[1-\cos 4(\alpha-\theta_i)\right]F_i(\beta_i)\right\},$$

$$X_{i8}=a\left\{F_i^{///}(\beta_i)\left[1-\cos 4(\alpha-\theta_i)\right]-4F_i^{//}(\beta_i)(\lambda-1)\sin 4(\alpha-\theta_i)\right.$$
$$+F_i^{/}(\beta_i)[3\lambda^2+1+\left(5\lambda^2-8\lambda-1\right)\cos 4(\alpha-\theta_i)]$$
$$\left.+2F_i(\beta_i)\left(\lambda^2-1\right)(\lambda-2)\sin 4(\alpha-\theta_i)\right\}-2b_1F_i^{///}(\beta_i)$$
$$-2F_i^{/}(\beta_i)[\lambda(\lambda+1)(a_{11}-a_{12}+b_1)$$
$$-(\lambda+1)(a_{11}-a_{12}-b_2)-2(a_{12}-a_{11})].$$

By substituting (14) into (18), we obtain a homogeneous system of linear algebraic equations for the constants A_i, B_i, C_i and D_i.

After some cumbersome calculations, from the existence condition for the nonzero solution of this system, we obtain the following characteristic equation for λ, which determines the stress concentration degree (6) see in (Galptshyan, 2008):

$$f(\lambda;a_{11},a_{12},a,\theta_1,\theta_2,\alpha)=0 \tag{19}$$

Equation (19) contains six independent parameters $a_{11}, a_{12}, a, \theta_1, \theta_2$ and α.

	a/b_1		a/b_1
Nb	− 0. 6423463	MgO	0. 2276457
CaF_2	− 0.4838456	Si	0. 2498694
FeS_2	− 0. 4066341	Ge	0. 275492
KCl	− 0. 2682469	Ta	0. 2874998
NaCl	− 0. 2154233	LiF	0. 3094264
V	− 0. 2139906	Fe	0. 4637442
Mo	− 0. 1877868	Ni	0. 4804368
TiC	− 0. 0664576	Ag	0. 5856406
W	0	Cu	0. 593247
Au	0. 0556095	Pb	0. 7026827
C	0. 0965294	Na	0. 8089901
Al	0. 1403437		

Table 1.

For certain specific values of these parameters, it follows from (6) and (13) that the stress components at the pole $r=0$ have an integrable singularity if $0<\mathrm{Re}\,\lambda<1$. In this case, the order of the singularity is equal to $|\mathrm{Re}\,\lambda-1|$.

Thus, studying the singularity of the stress state near the corner point of the interface between two crystals in the case of plane strain is reduced to finding the root of the transcendental equation (19) with the least positive real part.

A structural analysis of Eq. (19) shows that its left-hand side is a polynomial of degree 18 in a/b_1. The absolute value of a/b_1 is sufficiently small. Therefore, preserving only terms up to the first or the second degree in (19), instead of a polynomial of degree 18, we obtain a polynomial of the first or the second degree, i.e., various approximations to Eq. (19). We also note that for $a = 0$, from the above system of algebraic equations, just as from Eq. (19), we obtain the equation $\sin(\lambda+1)\pi = 0$ determining the eigenvalues $\lambda = \lambda_k = k \ (k \in \mathbf{N})$ which correspond to the plane strain of a homogeneous isotropic body.

Preserving only terms up to the first or second degree in a/b_1 in Eq. (19), we finally obtain

$$
\begin{aligned}
&2^8\left(\lambda^2-1\right)\{[\cos\lambda\alpha\cos\alpha-(\lambda+1)\cos\lambda(\alpha-\pi)\cos(\pi\lambda+\alpha)+\sin[(\alpha-\pi)(\lambda-1)]\\
&\times\sin\lambda\pi]\sin(\pi\lambda+\alpha)\sin\lambda(\alpha-\pi)+2\sin^2\lambda\pi\cos[(\lambda-1)(\alpha-\pi)]\cos[\alpha+\lambda(\pi-\alpha)]\}\\
&\times\cos^2[(\lambda+1)(\alpha-\pi)]\sin^4\lambda\pi+\frac{a}{b_1}\{2^4(\lambda-1)[\cos\lambda\alpha\cos\alpha-(\lambda+1)\cos\lambda(\alpha-\pi)\\
&\times\cos(\pi\lambda+\alpha)+\sin[(\alpha-\pi)(\lambda-1)]\sin\lambda\pi]\rho_{22}(\lambda)\cos^2[(\lambda+1)(\alpha-\pi)]\sin^4\lambda\pi\\
&-\rho_{21}(\lambda)\sin[(\lambda-1)(\alpha-\pi)]\sin\lambda\pi-\rho_{23}(\lambda)\sin[(\alpha-\pi)(\lambda-1)]\sin\lambda\pi+8(\lambda-1)\\
&\times[\cos\lambda\alpha\cos\alpha-(\lambda+1)\cos\lambda(\alpha-\pi)\cos(\pi\lambda+\alpha)+\sin[(\alpha-\pi)(\lambda-1)]\sin\lambda\pi]\\
&\times[\eta_3(5-3\lambda)\chi_1(\lambda)+(\lambda+1)\rho_9(\lambda)\chi_2(\lambda)-4\chi_{22}(\lambda)(\lambda+1)(1-\cos4\theta_1)]\chi_{11}(\lambda)\\
&\times\cos[(\lambda+1)(\alpha-\pi)]\sin^3\lambda\pi-\{2^4[\eta_3(\lambda-1)(3\lambda-5)\chi_2(\lambda)-\chi_3(\lambda)\rho_9(\lambda)\\
&-4(1-\cos4\theta_1)\chi_{31}(\lambda)](\lambda+1)\chi_{11}(\lambda)\sin^3\lambda\pi\cos[(\lambda+1)(\alpha-\pi)]+(\lambda-1)\{2^5(\lambda+1)\\
&\times\rho_1(\lambda)\sin^4\lambda\pi\cos[(\lambda+1)(\alpha-\pi)]\cos[(\lambda-1)(\alpha-\pi)]+\chi_{11}(\lambda)\{8\rho_3(\lambda)\sin\lambda\pi\\
&-\rho_{18}(\lambda)+4(\lambda-1)[\cos\alpha\cos\lambda\alpha-(\lambda+1)\cos\lambda(\alpha-\pi)\cos(\pi\lambda+\alpha)\\
&+\sin[(\alpha-\pi)(\lambda-1)]\sin\lambda\pi]\rho_{17}(\lambda)\cos[(\lambda+1)(\alpha-\pi)]\sin\lambda\pi\}\}+2^4(\lambda+1)\{\zeta_3(\lambda)\\
&\times[2-2\cos4(\alpha-\theta_1)+\rho_9(\lambda)]+(\lambda-1)[8\sin4(\alpha-\theta_1)+\eta_3(3\lambda-5)]\zeta_2(\lambda)-2\rho_4(\lambda)\\
&-4\chi_{44}(\lambda)(1-\cos4\theta_1)\}\chi_{11}(\lambda)\sin^3\lambda\pi\cos[(\lambda+1)(\alpha-\pi)]+\{2^5(\lambda+1)\rho_1(\lambda)\\
&\times\sin^4\lambda\pi\cos[(\lambda+1)(\alpha-\pi)]\cos[(\lambda-1)(\alpha-\pi)]+[8(\lambda+1)\rho_{11}(\lambda)\sin\lambda\pi-\rho_{18}(\lambda)]\\
&\times\chi_{11}(\lambda)\}\}\cos[(\lambda-1)(\alpha-\pi)]\sin\lambda\pi\}=0,
\end{aligned}
\tag{20}
$$

$$\eta_1=\sin4(\alpha-\theta_2)-\sin4(\alpha-\theta_1),$$

$$\eta_2=\cos4(\alpha-\theta_2)-\cos4(\alpha-\theta_1),\quad \eta_3=\sin4\theta_2-\sin4\theta_1,\quad \eta_4=\cos4\theta_2-\cos4\theta_1,$$

$$\zeta_1(\lambda)=2(\lambda\cos\alpha\sin\lambda\alpha-\sin\alpha\cos\lambda\alpha),\quad \zeta_2(\lambda)=2\cos\alpha\cos\lambda\alpha,$$

$$\zeta_3(\lambda)=2(\lambda\sin\lambda\alpha\cos\alpha+\sin\alpha\cos\lambda\alpha),\quad \chi_{11}(\lambda)=2\sin\lambda\pi\cos[(\lambda+1)(\alpha-\pi)],$$

$$\chi_1(\lambda)=(\lambda+1)\sin(\lambda-1)\alpha-(\lambda-1)\sin(\lambda+1)\alpha,\quad \chi_2(\lambda)=2\sin\alpha\sin\lambda\alpha,$$

$$\chi_{22}(\lambda)=2\sin(\pi\lambda+\alpha)\sin(\alpha-\pi)\lambda,\quad \chi_3(\lambda)=2(\lambda\sin\alpha\cos\lambda\alpha+\sin\lambda\alpha\cos\alpha),$$

$$\chi_{31}(\lambda)=(\lambda-1)\sin\left[(\lambda-1)(\alpha-2\pi)\right]-(\lambda+1)\sin(\lambda+1)\alpha,$$

$$\chi_{44}(\lambda)=(\lambda-1)\sin\left[(\lambda-1)(\alpha-2\pi)\right]+(\lambda+1)\sin(\lambda+1)\alpha,$$

$$\chi_{33}(\lambda)=2\cos\lambda(\alpha-\pi)\cos(\pi\lambda+\alpha),\quad \chi_{12}(\lambda)=-2\sin\lambda\pi\sin\left[(\lambda+1)(\alpha-\pi)\right],$$

$$\chi_{13}(\lambda)=(\lambda-1)\sin(\lambda+1)\alpha+(\lambda+1)\sin\left[(\lambda-1)(\alpha-2\pi)\right],$$

$$\chi_{21}(\lambda)=(\lambda+1)\sin\left[(\lambda-1)(\alpha-2\pi)\right]-(\lambda-1)\sin(\lambda+1)\alpha,$$

$$\rho_9(\lambda)=\eta_4(\lambda-1)+4(1-\cos 4\theta_1),$$

$$\rho_1(\lambda)=3\eta_4\chi_1(\lambda)+\eta_3\chi_2(\lambda)(\lambda+1)-4\chi_{11}(\lambda)(1-\cos 4\theta_1),$$

$$\begin{aligned}\rho_2(\lambda)=&\left[3\eta_2\lambda+4\cos 4(\alpha-\theta_1)-3\cos 4(\alpha-\theta_2)-1\right](\lambda+1)\sin\left[(\lambda+1)(\alpha-2\pi)\right]\\&-\left[3\eta_1\lambda+\sin 4(\alpha-\theta_1)+3\sin 4(\alpha-\theta_2)\right](\lambda-1)\cos\left[(\lambda+1)(\alpha-2\pi)\right]\\&+\left[1-\cos 4(\alpha-\theta_1)\right](\lambda+1)\sin(\lambda+1)\alpha+4(\lambda-1)\sin 4(\alpha-\theta_1)\cos(\lambda+1)\alpha,\end{aligned}$$

$$\rho_{13}(\lambda)=(3\lambda-5)\eta_4+4(1-\cos 4\theta_1),$$

$$\begin{aligned}\rho_{14}(\lambda)=&\left[\lambda\cos 4(\alpha-\theta_1)-(\lambda-1)\cos 4(\alpha-\theta_2)-1\right](\lambda+1)\sin\left[(\lambda-1)(\alpha-2\pi)\right]\\&-\left(1-\cos 4(\alpha-\theta_1)\right)(\lambda-1)\sin(\lambda+1)\alpha+\eta_1\left(\lambda^2-1\right)\cos\left[(\lambda-1)(\alpha-2\pi)\right],\end{aligned}$$

$$\begin{aligned}\rho_{16}(\lambda)=&\left(1+\lambda\cos 4(\alpha-\theta_1)-(\lambda+1)\cos 4(\alpha-\theta_2)\right)\cos\left[(\lambda+1)(\alpha-2\pi)\right]\\&-\left(1-\cos 4(\alpha-\theta_1)\right)\cos(\lambda+1)\alpha-\eta_1(\lambda+1)\sin\left[(\lambda+1)(\alpha-2\pi)\right],\end{aligned}$$

$$\begin{aligned}\rho_6(\lambda)=&\left(1+\lambda\cos 4(\alpha-\theta_1)-(\lambda+1)\cos 4(\alpha-\theta_2)\right)\sin\left[(\lambda+1)(\alpha-2\pi)\right]\\&-\left(1-\cos 4(\alpha-\theta_1)\right)\sin(\lambda+1)\alpha+\eta_1(\lambda+1)\cos\left[(\lambda+1)(\alpha-2\pi)\right],\end{aligned}$$

$$\rho_7(\lambda)=3\eta_4\zeta_1(\lambda)+\eta_3(\lambda+1)\zeta_2(\lambda)+2\rho_6(\lambda)+4\chi_{11}(\lambda)(1-\cos 4\theta_1),$$

$$\begin{aligned}\rho_8(\lambda)=&\left[\lambda\cos 4(\alpha-\theta_1)-(\lambda-1)\cos 4(\alpha-\theta_2)-1\right]\cos\left[(\lambda-1)(\alpha-2\pi)\right]\\&-\left[1-\cos 4(\alpha-\theta_1)\right]\cos(\lambda+1)\alpha-\eta_1(\lambda-1)\sin\left[(\lambda-1)(\alpha-2\pi)\right],\end{aligned}$$

$$\begin{aligned}\rho_{19}(\lambda)=&\left(\lambda^2-1\right)\Big[\left(3\eta_2\lambda+4\cos 4(\alpha-\theta_1)-5\cos 4(\alpha-\theta_2)+1\right)\cos\left[(\lambda-1)(\alpha-2\pi)\right]\\&+\left(3\eta_1\lambda+\sin 4(\alpha-\theta_1)-5\sin 4(\alpha-\theta_2)\right)\sin\left[(\lambda-1)(\alpha-2\pi)\right]\\&+\left(1-\cos 4(\alpha-\theta_1)\right)\cos(\lambda+1)\alpha\Big]-4(\lambda-1)^2\sin 4(\alpha-\theta_1)\sin(\lambda+1)\alpha,\end{aligned}$$

$$\rho(\lambda)=\left(3\eta_2\lambda+4\cos 4(\alpha-\theta_1)-3\cos 4(\alpha-\theta_2)-1\right)(\lambda+1)\cos\left[(\lambda+1)(\alpha-2\pi)\right]$$
$$+\left(3\eta_1\lambda+\sin 4(\alpha-\theta_1)+3\sin 4(\alpha-\theta_2)\right)(\lambda-1)\sin\left[(\lambda+1)(\alpha-2\pi)\right]$$
$$+\left(1-\cos 4(\alpha-\theta_1)\right)(\lambda+1)\cos(\lambda+1)\alpha-4(\lambda-1)\sin 4(\alpha-\theta_1)\sin(\lambda+1)\alpha,$$

$$\rho_0(\lambda)=\eta_3(\lambda+1)\zeta_3(\lambda)-3\eta_4\left(\lambda^2-1\right)\zeta_2(\lambda)+2\rho(\lambda)-4(\lambda+1)(1-\cos 4\theta_1)\chi_{12}(\lambda),$$

$$\rho_5(\lambda)=3\eta_4\chi_2(\lambda)(\lambda-1)+\eta_3\chi_3(\lambda)-4(1-\cos 4\theta_1)\chi_{12}(\lambda),$$

$$\rho_3(\lambda)=\chi_{11}(\lambda)\left[3\eta_3\zeta_2(\lambda)\left(\lambda^2-1\right)+\eta_4(\lambda+1)\zeta_3(\lambda)-2\rho_2(\lambda)\right.$$
$$\left.+4(1-\cos 4\theta_1)(\lambda+1)\chi_{11}(\lambda)-(\lambda+1)\rho_1(\lambda)\right]$$
$$+\chi_{12}(\lambda)\left[3\eta_3(\lambda+1)\chi_1(\lambda)-\eta_4(\lambda+1)^2\chi_2(\lambda)\right.$$
$$\left.+4(\lambda+1)(1-\cos 4\theta_1)\chi_{12}(\lambda)-\rho_0(\lambda)\right],$$

$$\rho_{11}(\lambda)=\eta_4\left[\chi_{11}(\lambda)\chi_3(\lambda)-\chi_{12}(\lambda)\chi_2(\lambda)(\lambda+1)\right]-3\eta_3\left[\chi_{11}(\lambda)\chi_2(\lambda)(\lambda-1)\right.$$
$$\left.-\chi_1(\lambda)\chi_{12}(\lambda)\right]+2^4(1-\cos 4\theta_1)\sin^2\lambda\pi-\rho_1(\lambda)\chi_{11}(\lambda)-\rho_5(\lambda)\chi_{12}(\lambda),$$

$$\rho_{12}(\lambda)=\rho_{11}(\lambda)\chi_{11}(\lambda)-4\rho_1(\lambda)\sin^2\lambda\pi,$$

$$\rho_{15}(\lambda)=4\rho_7(\lambda)\sin^2\lambda\pi+\rho_{11}(\lambda)\chi_{11}(\lambda),$$

$$\rho_{17}(\lambda)=2(\lambda+1)\left(\eta_4\chi_{11}(\lambda)-\eta_3\chi_{12}(\lambda)\right)\cos(\lambda-1)\alpha-6(\lambda+1)$$

$$\times\left(\eta_3\chi_{11}(\lambda)+\eta_4\chi_{12}(\lambda)\right)\sin(\lambda-1)\alpha+2\left(\chi_{11}(\lambda)\rho_{16}(\lambda)-\chi_{12}(\lambda)\rho_6(\lambda)\right),$$

$$\rho_{18}(\lambda)=-4\chi_{11}(\lambda)\left\{\left[\zeta_1(\lambda)\left(2-2\cos 4(\alpha-\theta_1)+\rho_{13}(\lambda)\right)+\eta_3\zeta_2(\lambda)\left(\lambda^2-1\right)\right.\right.$$
$$\left.+2\rho_{14}(\lambda)-4\chi_{13}(\lambda)(1-\cos 4\theta_1)\right]\sin^2(\lambda+1)\pi+\rho_{13}(\lambda)\chi_1(\lambda)+\eta_3\chi_2(\lambda)\left(\lambda^2-1\right)$$
$$\left.-4\chi_{21}(\lambda)(1-\cos 4\theta_1)\right\}+2(\lambda+1)\rho_{12}(\lambda)\sin\lambda\pi\cos\left[(\lambda-1)(\alpha-\pi)\right]$$

$$+2(\lambda+1)\left[\rho_{15}(\lambda)+\rho_{17}(\lambda)\chi_{12}(\lambda)\right]\sin\lambda\pi\cos\left[(\alpha-\pi)(\lambda-1)\right]$$
$$+2\rho_{17}(\lambda)\chi_{11}(\lambda)(\lambda-1)\sin\lambda\pi\sin\left[(\alpha-\pi)(\lambda-1)\right],$$

$$\rho_4(\lambda)=(\lambda-1)\left[\left(3\eta_2\lambda+4\cos 4(\alpha-\theta_1)-5\cos 4(\alpha-\theta_2)+1\right)\sin\left[(\lambda-1)(\alpha-2\pi)\right]\right.$$
$$-\left(3\eta_1\lambda+\sin 4(\alpha-\theta_1)-5\sin 4(\alpha-\theta_2)\right)\cos\left[(\lambda-1)(\alpha-2\pi)\right]+4\sin 4(\alpha-\theta_1)$$
$$\left.\times\cos(\lambda+1)\alpha\right]+(\lambda+1)\left[1-\cos 4(\alpha-\theta_1)\right]\sin(\lambda+1)\alpha$$

$$\rho_{20}(\lambda)=-4\chi_{11}(\lambda)\big[\eta_3\zeta_3(\lambda)(\lambda^2-1)-(2-2\cos4(\alpha-\theta_1)+\rho_{13}(\lambda))(\lambda^2-1)\zeta_2(\lambda)$$
$$+8(\lambda-1)\zeta_1(\lambda)\sin4(\alpha-\theta_1)+2(\lambda+1)(\rho_{19}(\lambda)$$
$$+2\chi_{33}(\lambda)(1-\cos4\theta_1)(\lambda^2-1))\big]\sin^2\lambda\pi+\rho_{12}(\lambda)(\lambda^2-1)\big[\zeta_2(\lambda)-\chi_{33}(\lambda)(\lambda+1)\big]$$
$$+4(\lambda+1)\chi_{12}(\lambda)\big[\rho_{13}(\lambda)\chi_1(\lambda)+\eta_3\chi_2(\lambda)(\lambda^2-1)-4\chi_{21}(\lambda)(1-\cos4\theta_1)\big]\sin^2\lambda\pi$$
$$-2\big[\rho_{11}(\lambda)\chi_{12}(\lambda)(\lambda+1)-4\rho_0(\lambda)\sin^2\lambda\pi\big](\lambda+1)\sin\lambda\pi\cos\big[(\alpha-\pi)(\lambda-1)\big]$$

$$+4\chi_{11}(\lambda)(\lambda^2-1)\big[\rho_{13}(\lambda)\chi_2(\lambda)+\eta_3\chi_3(\lambda)$$
$$-4\chi_{22}(\lambda)(1-\cos4\theta_1)\big]\sin^2\lambda\pi+2(\lambda-1)\big[\rho_3(\lambda)\chi_{11}(\lambda)$$
$$-4\rho_1(\lambda)(\lambda+1)\sin^2\lambda\pi\big]\sin\lambda\pi\sin\big[(\alpha-\pi)(\lambda-1)\big]$$
$$-4\chi_{12}(\lambda)(\lambda+1)\big[\rho_{13}(\lambda)\chi_1(\lambda)+\eta_3\chi_2(\lambda)(\lambda^2-1)$$
$$-4\chi_{21}(\lambda)(1-\cos4\theta_1)\big]\sin^2\lambda\pi$$
$$+2(\lambda+1)\big[\rho_3(\lambda)\chi_{12}(\lambda)-\rho_5(\lambda)(\lambda+1)\big]\sin\lambda\pi\cos\big[(\alpha-\pi)(\lambda-1)\big],$$

$$\rho_{22}(\lambda)=\eta_3\zeta_1(\lambda)(5-3\lambda)+\zeta_2(\lambda)(\lambda+1)\big[2-2\cos4(\alpha-\theta_1)+\rho_9(\lambda)\big]$$
$$+2(\lambda+1)\big[\rho_8(\lambda)-2\chi_{33}(\lambda)(1-\cos4\theta_1)\big],$$

$$\rho_{23}(\lambda)=4(1-\lambda^2)\big[\cos\alpha\cos\lambda\alpha-(\lambda+1)\cos\lambda(\alpha-\pi)\cos(\pi\lambda+\alpha)$$
$$+\sin\big[(\alpha-\pi)(\lambda-1)\big]\sin\lambda\pi\big](\rho_{17}(\lambda)\chi_{12}(\lambda)+\rho_{15}(\lambda))\cos\big[(\lambda+1)(\alpha-\pi)\big]\sin\lambda\pi$$
$$-\rho_{20}(\lambda)\chi_{11}(\lambda)-8(\lambda+1)\big[\rho_{11}(\lambda)\chi_{12}(\lambda)(\lambda+1)-4\rho_0(\lambda)\sin^2\lambda\pi$$
$$-\rho_3(\lambda)\chi_{12}(\lambda)+\rho_5(\lambda)(\lambda+1)\big]\sin^2\lambda\pi\cos\big[(\lambda+1)(\alpha-\pi)\big]\cos\big[(\lambda-1)(\alpha-\pi)\big]$$

$$\rho_{21}(\lambda)=4(1-\lambda^2)\big[\cos\alpha\cos\lambda\alpha-(\lambda+1)\cos\lambda(\alpha-\pi)\cos(\pi\lambda+\alpha)$$
$$+\sin\big[(\alpha-\pi)(\lambda-1)\big]\sin\lambda\pi\big]\big[\chi_{11}(\lambda)\rho_{11}(\lambda)-4\rho_1(\lambda)\sin^2\lambda\pi\big]$$
$$\times\cos\big[(\lambda+1)(\alpha-\pi)\big]\sin\lambda\pi-\chi_{11}(\lambda)\rho_{20}(\lambda).$$

6. Study of the roots of the characteristic equation

Table 1 shows the values of the dimensionless ratio a/b_1 for some cubic crystals at room temperature. Moreover for all the cosidered materials $b_1>0$ and with the exception of cubic pyrits (FeS_2), for which $b_2=0,00365798\cdot10^{-11}\mathrm{Pa}^{-1}$, $b_2<0$.

The least value of the ratio is attained for the niobium crystal (Nb) and the largest, for the sodium crystal (Na). In absolute value, $|a/b_1|<1$.

To study the roots of Eq. (19) in the interval $10<\mathrm{Re}\,\lambda<1$, in Table 1 we choose six real materials and two imaginary materials for which $|a/b_1|=10^{-5}$. To investigate whether there is a singularity in the stress concentration at the corner point of the interface between the two joined crystals, for each of the materials, we choose seven versions of variations in the parameters α, θ_1 and θ_2, which are given in Tables 2 and 3. For example, the first

	$\frac{a}{b_1}$	σ_*, *МПа*	$\alpha=\pi/2$ $\theta_1=\pi/4$ $\theta_2=0$	$\alpha=\pi/4$ $\theta_1=\pi/4$ $\theta_2=0$
Mo	−0.1877868	800 − 1200	0.647029	0.0174393 $\mp i\cdot 0.058343$ 0.688156
TiC	− 0.0664576	560	0.0153193 0.6899690	0.01012946 $\mp i\cdot 0.047990$ 0.72254
	-10^{-5}		$7.3815\cdot 10^{-6}$ 0. 987524	0.996981
W	0	1100 1800 − 4150	-	-
	10^{-5}		0.994154 $\mp i\cdot 0.0107712$	0.000786231 0.9276061 $\mp i\cdot 0.1747073$
Au	0.0556095	140	0.0497266 0.422350	0.0809312 0.2043483 0.4714287 0.9546085 $\mp i\cdot 0.216914$
C	0.0965294		0.032592 0.5279915	0.5889015
Al	0.1403437	50 115	0.0284796 0.560425	0.0522492 $\mp i\cdot 0.073474$ 0.617351

Table 2.

version, where $\alpha = \pi / 2$, $\theta_1 = \pi / 4$, $\theta_2 = 0$, concerns the case in which the interface between two crystals is formed by the plane of elastic symmetry of the second crystal but not of the first crystal. In the fourth version $(\alpha = \pi/4, \theta_1 = 0, \theta_2 = \pi/4)$, the part $\varphi_1 = \theta_1 = 0$ of the interface is the plane of elastic symmetry of the first crystal, and the other part $(\varphi = \theta_2 = \pi/4)$ is the plane of elastic symmetry of the second crystal.

For all materials given in Tables 2 and 3 and for all versions, we found, in general, all realand complex roots of Eq. (20) with $0 < \mathrm{Re}\,\lambda < 1$, including all (without any exception) rootswith minimum positive real part.

It follows from Tables 2 and 3 that, for all two-crystals except tungsten and for all the versions, there are stress concentrations near the corner point of the interface between the crystals. If we compare the two crystals of molybdenum (Mo) and titanium carbide (TiC) for which $a/b_1 < 0$, then it follows from the results obtained for seven versions that, in general, the stress concentration degree (the order of singularity) of molybdenum is less than that of titanium carbide. It is of interest to note that the ultimate strength of polycrystalline molybdenum σ_* is larger than the ultimate strength of polycrystalline titanium carbide, which is an integral characteristic of strength. In Table 2, we present the ultimate strengths under tension at temperature 20^0C for molybdenum and titanium carbide.

For the two-crystal of tungsten (W), we have $a/b_1 = 0$ and hence, according to (20), there is no singularity of stress concentration near the corner point of the interface between two crystals. This may be one of the causes of the fact that the polycrystalline tungsten materials have very high ultimate strength.

In Table 2, we present the ultimate strengths under tension of the polycrystalline tungsten annealed wire (1100 MPa) and unannealed wire (from 1800 MPa to 4150 MPa, depending on the diameter). We draw the reader's attention to the fact that the ultimate strength of the diamond monocrystal at temperature $20^\circ C$ is equal to 1800 MPa.

Note that for the polycrystalline metals listed in Table 2 there is a correspondence between the ultimate strength σ_* and the modulus of elasticity E (here the quantity E is treated as an integral characteristic of elasticity of a metal). The moduli of elasticity of the polycrystalline metals Mo, W, Au and Al listed in Table 2 are, respectively, equal to (285-300) GPa, (350-380) GPa, 79 GPa, and 70 GPa. The ultimate strength is larger for a metal with larger modulus of elasticity.

All numerical values of strength limit brought in the table (2) as well as elastic modulus for the discussed materials considered to be a published data taken from various sources. For example, these data for tungsten (W) are taken from the book (Knuniants and etc. 1961).

Strength limit of unannealed tungsten wire is depended from the diameter and could be explained by the existence defects of crystal lattice.

Here we also note that there is no such correspondence if molybdenum and titanium carbide are compared. Although the ultimate strength of molybdenum is larger than the ultimate strength of titanium carbide, the modulus of elasticity of molybdenum is less than the modulus of elasticity of titanium carbide, which is equal to 460 GPa. We note that the titanium carbide is a compound matter.

	$\alpha=\pi/2$ $\theta_1=0$ $\theta_2=\pi/4$	$\alpha=\pi/4$ $\theta_1=0$ $\theta_2=\pi/4$	$\alpha=3\pi/4$ $\theta_1=\pi/4$ $\theta_2=0$	$\alpha=3\pi/4$ $\theta_1=\pi/6$ $\theta_2=\pi/3$	$\alpha=\pi/2$ $\theta_1=\pi/6$ $\theta_2=\pi/3$
Mo	0.710072	0.0775947	0.0411225	0.957018	0.0094110 $\mp i\cdot 0.0739984$ 0.656204
TiC	0.774834	0.0542824	0.0365724	0.95741	0.0071750 $\mp i\cdot 0.064893$ 0.702898
	0.993177	$7.87311\cdot 10^{-4}$ 0.927667 $\mp i\cdot 0.1746807$ 0.999978 $\mp i\cdot 0.0029137$	$8.82199\cdot 10^{-4}$ 0.9905998 $\mp i\cdot 0.1403189$ 0.995236	0.9906207 $\mp i\cdot 0.1403553$ 0.995563	0.9960624 $\mp i\cdot 0.0057440$
W	-	-	-	-	-
	$1.0567\cdot 10^{-5}$ 0.9970875 $\mp i\cdot 0.00590$	0.9276334 $\mp i\cdot 0.174718$ 0.997109	0.9912171 $\mp i\cdot 0.140329$ 0.99971 $\mp i\cdot 0.005039$	0.0011696 0.991197 $\mp i\cdot 0.1402922$ 0.999734 $\mp i\cdot 0.004690$	0.0013220 0.993404
Au	0.0644867 0.8890054 $\mp i\cdot 0.4002118$	0.8730987 $\mp i\cdot 0.236169$	0.0779069 0.2491514 0.7677329	0.1780295 0.7444930	0.4007251
C	0.1243433	0.8400085 $\mp i\cdot 0.272280$	0.0557915 0.4864447 0.7224011	0.332059 0.6977614	0.5246499
Al	0.215732	0.0206982 $\mp i.\cdot 0.113575$	0.0512975 0.612502 0.65982	0.451447 0.655004	0.0154580 $\mp i\cdot 0.094053$ 0.56112

Table 3.

Discussing the results obtained for two-crystals of gold (Au) and aluminum (Al) (Tables 2 and 3), for which $a/b_1 > 0$, we conclude that, according to the root of Eq. (20) obtained for seven versions, the stress concentration degree (the order of singularity) near the corner point of the interface between two crystals is larger for the two-crystal of aluminum. Here we also note that the ultimate strength and the modulus of elasticity of polycrystalline gold are larger than those of polycrystalline aluminum. In Table 2, we present the ultimate strengths under tension for polycrystalline aluminum annealed wire (50 MPa) and cold-rolled wire (115 MPa).

For a two-crystal of diamond (C), the stress concentrations near the corner point of the interface between two crystals are rather large (see Tables 2 and 3).

Depending on the choice of the coordinate axes, the modulus of elasticity of the diamond monocrystal varies from 1049.67 GPa to 1206.63 GPa, and, as was already noted, the ultimate strength is approximately equal to 1800 MPa. But for diamond polycrystalline formations (edge, aggregate), we did not found the corresponding integral characteristics of elasticity and strength in the literature. We assume that these characteristics, numerically, must be less than the modulus of elasticity and the ultimate strength of the diamond monocrystal, because there is no stress concentration in the interior of a polycrystalline body.

As follows from Tables 2 and 3, for the imaginary materials with the ratios $|a/b_1| = 10^{-5}$, there are very strong stress concentrations for some of the versions.

In Figs. 2–5, we present graphs of variation of the function $r_*^{\mathrm{Re}\,\lambda-1}$ as r_* approaches the pole $r = 0$

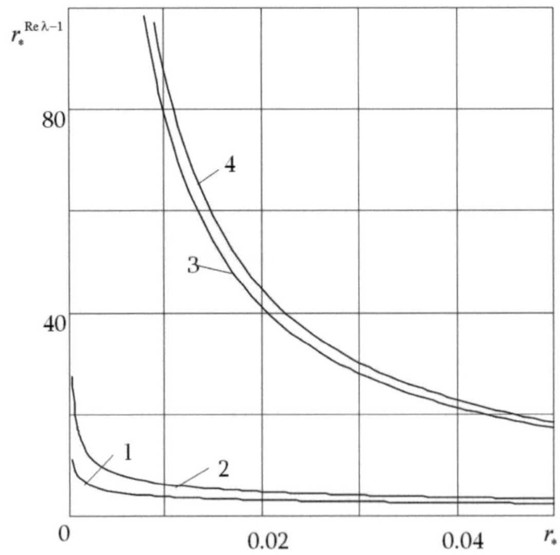

Fig. 2.

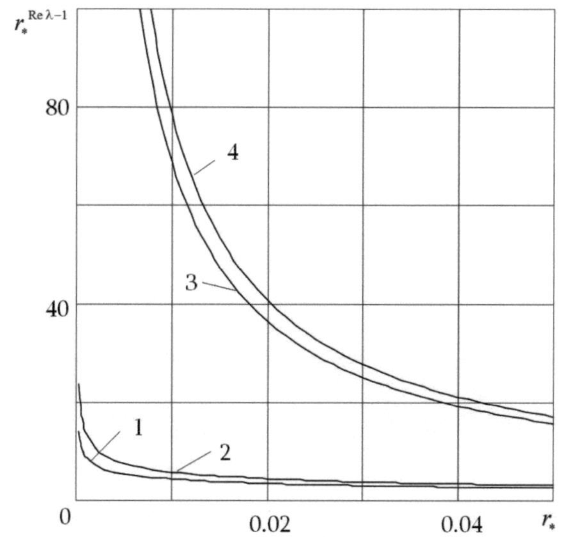

Fig. 3.

(r_* is the ratio of the coordinate r to the characteristic dimension of the two-crystal). Curves 3 and 4 correspond to the two-crystal of gold (Au) and the two-crystal of aluminum (Al), respectively. Curves 1 and 2 correspond to a two-component piecewise homogeneous isotropic body with shear moduli ratio $\mu = G_1 / G_2 = a_{44}^{(2)} / a_{44}^{(1)} = 20$ and Poisson ratios $\nu_1 = 0.2,\ \nu_2 = 0.4$ and to the two-component piecewise homogeneous isotropic body with shear moduli ratio $\mu = G_1 / G_2 = a_{44}^{(2)} / a_{44}^{(1)} = 0.05$ and the Poisson ratios $\nu_1 = 0.2,\ \nu_2 = 0.3$, respectively. Moreover, $\nu_1 = -E_1 a_{12}^{(1)},\ \nu_2 = -E_2 a_{12}^{(2)}$,

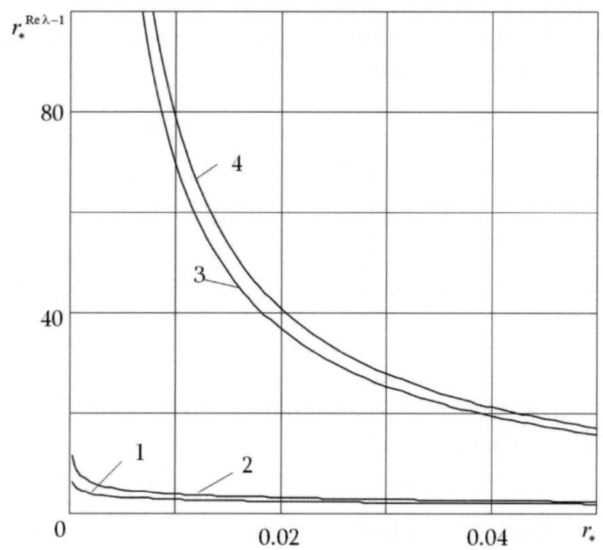

Fig. 4.

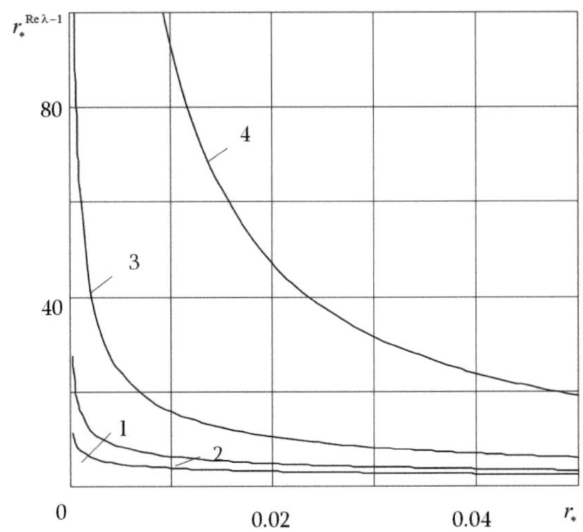

Fig. 5.

where $a_{12}^{(1)}$ and $a_{12}^{(2)}$ are the strain coefficients of homogeneous isotropic parts and E_1 and E_2 are the Young moduli of the same homogeneous isotropic parts. Figures 2-5 correspond to the first, second, fifth, and seventh versions given in Tables 2 and 3, respectively; curves 1 and 2 in the same figures correspond to the four values of the linear angle α formed by the contact surfaces of homogeneous isotropic parts of the compound body. The values of the angle α in Figs. 2–5 are respectively equal to: $\alpha = \pi/2, \pi/4, 3\pi/4$ and $\pi/2$. The values of the ratio μ and the Poisson ratios ν_1 and ν_2, and the corresponding values of the orders of singularities, are taken from Table 1 presented in (Chobanyan, 1987; Chobanyan & Gevorkyan, 1971).

The graphs show that the order of singularity of the stresses at the corner point of the contact surface of aluminum crystals is larger than the order of singularity of stresses at the corner point of the contact surface of gold crystals. The graphs also show that, for the piecewise homogeneous isotropic bodies under study, the order of singularity of the stresses is much lower than that for two-crystals of aluminum and gold.

7. Conclusion

From the analysis performed in Section 6, we draw the following conclusions.

Although we considered specific cases of stress state, namely, the out-of-plane strain and the plane strain of two-crystals whose separate crystals consist of one and the same material with cubic symmetry and with different orientations of the principal directions of elasticity, we can state that, in the general case of loading of a polycrystalline body, there are stress concentrations at the corner points of the interface between the joined crystals.

It is well known that the structure of the crystal lattice of a given matter plays a definite role in the process of formation of its mechanical properties and characteristics, in particular, the strength of monocrystals. But in polycrystalline materials, along with this factor, the strength of the joint of crystals and the fact that there are stress singularities at the corner points of the interface between the crystals totally play the decisive role in the process of formation of these characteristics. This can be observed in the process of mechanical fragmentation of polycrystalline materials. They split and form small crystals of certain shape. Of course, the separate crystals are also deformed in this process. The modulus of elasticity and the ultimate strength of a monocrystal with cubic symmetry for simple matters is larger than the corresponding characteristics of the polycrystalline material of the same matter.

In the problem of plane strain, the existence of stress concentration (singularity) at the corner point of the interface between the two joined crystals with cubic symmetry made of the same material, just as the degree of stress concentration (the order of singularity), depends on the parameters a/b_1, α, θ_1, and θ_2, which are determined in Sections 1-4.

In the case of out-of-plane strain of the two-crystal under study, there is no stress concentration at the corner point of the interface between the two joined crystals.

8. References

Alexandrov V. M. and Romalis B. L., *Contact Problems in Mechanical Engineering* (Mashinostroenie, Moscow, 1986) [in Russian].

Alexandrov V. M. and Pozharskii D. A., *Nonclassical Spatial Problems in Mechanics of Contact Interactions between Elastic Bodies* (Faktorial,Moscow, 1998) [in Russian].

Alexandrov V. M. and Pozharskii D. A., "On the 3D Contact Problem for an Elastic Cone with Unknown Contact Area," Izv. Akad.Nauk.Mekh. Tverd. Tela, No. 2, 36–41 (1998) [Mech. Solids (Engl. Transl.) 33 (2), 29–34 (1998)].

Alexandrov V.M., Kalker D. D., and Pozharskii D. A., "Calculation of Stresses in the Axisymmetric Contact Problem for a Two-Layered Elastic Base," Izv.Akad. Nauk.Mekh. Tverd. Tela,No. 5, 118–130 (2000) [Mech. Solids (Engl. Transl.) 35 (5), 97–106 (2000)].

Alexandrov V. M. and Klindukhov V. V., "Contact Problems for a Two-Layer Elastic Foundation with a Nonideal Mechanical Constraint between the Layers," Izv. Akad. Nauk. Mekh. Tverd. Tela, No. 3, 84–92 (2000) [Mech. Solids (Engl. Transl.) 35 (3), 71–78 (2000)].

Aleksandrov V. M., "Doubly Periodic Contact Problems for an Elastic Layer," Prikl. Mat. Mekh. 66 (2), 307-315 (2002) [J. Appl.Math. Mech. (Engl. Transl.) 66 (2), 297–305 (2002)].

Aleksandrov V. M. and Pozharskii D. A., "Three-Dimensional Contact Problems Taking Friction and Non-Linear Roughness into Account," Prikl. Mat. Mekh. 68 (3), 516-526 (2004) [J. Appl. Math. Mech. (Engl. Transl.) 68 (3), 463–472 (2004)].

Alexandrov V.M. and Kalyakin A. A., "Plane and Axisymmetric Contact Problems for a Three-Layer Elastic Half-Space," Izv. Akad. Nauk.Mekh. Tverd. Tela, No. 5, 30-38 (2005) [Mech. Solids (Engl. Transl.) 40 (5), 20–26 (2005)].

Alexandrov V.M. and Klindukhov V. V., "An Axisymmetric Contact Problem for Half-Space Inhomogeneous in Depth," Izv. Akad. Nauk. Mekh. Tverd. Tela, No. 2, 55–60 (2005) [Mech. Solids (Engl. Transl.) 40 (2), 46–50 (2005)].

Alexandrov V. M., "Longitudinal Crack in an Orthotropic Elastic Strip with Free Faces," Izv. Akad. Nauk. Mekh. Tverd. Tela,No. 1, 115–124 (2006) [Mech. Solids (Engl. Transl.) 41 (1), 88–94 (2006)].

Aleksandrov V.M., "Two Problems with Mixed Boundary Conditions for an Elastic Orthotropic Strip," Prikl. Mat. Mekh. 70 (1), 139–149 (2006) [J. Appl.Math. Mech. (Engl. Transl.) 70 (1), 128–138 (2006)].

Belubekyan V. M., "Is there a Singularity at a Corner Point of Crystal Junction?" in *Investigations of Contemporary Scientific Problems in Higher Educational Institutions* (Aiastan, Erevan, 2000), pp. 139- 143.

Chobanyan K. S. and Gevorkyan S. Kh., "Stress Field Behavior near a Corner Point of the Interface in the Problem of Plane Strain of a Compound Elastic Body," Izv. Akad. Nauk Armyan. SSR. Ser. Mekh. 24 (5), 16–24 (1971).

Chobanyan K. S., *Stresses in Compound Elastic Bodies* (Izd-vo Akad. Nauk Armyan. SSR, Erevan, 1987).

Feodos'ev V. I., *Strength of Materials* (Nauka, Moscow, 1979) [in Russian].

Galptshyan P.V., "On the Existence of Stress Concentrations in Loaded Bodies Made of Polycrystalline Materials," Izv. Akad. Nauk. Mekh. Tverd. Tela, No 6, 149-166 (2008) [Mech. Solids (Engl. Transl.) 43(6), 967-981 (2008)].

Knuniants I. L. et al. (Editors), Short chemical encyclopedia, Vol. 1 (Sovietskaya encyclopedia, Moscow, 1961) [in Russian].

Lekhnitskii S. G., *Theory of Elasticity of an Anisotropic Body* (Nauka, Moscow, 1977; Mir Publishers, Moscow, 1981).

Love A. E. H., *A Treatise on the Mathematical Theory of Elasticity*, 4th ed. (Cambridge Univ. Press, Cambridge, 1927; ONTI,Moscow, 1935).

Osrtrik V. I. and Ulitko A. F., "Contact between Two ElasticWedges with Friction," Izv. Akad. Nauk.Mekh.Tverd.Tela, No. 3, 93–100 (2000) [Mech. Solids (Engl. Transl.) 35 (3), 79–85 (2000)].

Pozharskii D. A. and Chebakov M. I., "On Singularities of Contact Stresses in the Problem of a Wedge-Shaped Punch on an ElasticCone," Izv. Akad. Nauk.Mekh. Tverd. Tela, No. 5, 72–77 (1998) [Mech. Solids (Engl. Transl.) 33 (5), 57–61 (1998)].

Pozharskii D. A., "The Three-Dimensional Contact Problem for an Elastic Wedge Taking Friction Forces into Account," Prikl. Mat. Mekh. 64 (1), 151–159 (2000) [J. Appl. Math. Mech. (Engl. Transl.) 64 (1), 147–154 (2000)].

Pozharskii D. A., "Contact with Adhesion between Flexible Plates and an ElasticWedge," Izv. Akad. Nauk. Mekh. Tverd. Tela,No. 4, 58–68 (2004) [Mech. Solids (Engl. Transl.) 39 (4), 46–54 (2004)].

Timoshenko S. P. and Goodyear J. N., *Theory of Elasticity* (McGraw-Hill, New York, 1951; Nauka, Moscow, 1975).

Ulitko A. F. and Kochalovskaya N. E., "Contact Interaction between a Rigid and Elastic Wedges at Initial Point Contact at Their Common Vertex," Dokl. Nats. Akad. Nauk Ukrainy. Ser. Mat. Estestvozn., Tekhn.N., No. 1, 51–54 (1995).

Vainstein B. K. et al. (Editors), *Modern Crystallography*, Vol. 4 (Nauka, Moscow, 1981) [in Russian].

3

Grain-Scale Modeling Approaches for Polycrystalline Aggregates

Igor Simonovski[1] and Leon Cizelj[2]
[1]*European Commission, DG-JRC, Institute for Energy and Transport, P.O. Box 2, NL-1755 ZG Petten*
[2]*Jožef Stefan Institute, Reactor Engineering Division, Jamova cesta 39, SI-1000 Ljubljana*
[1]*The Netherlands*
[2]*Slovenia*

1. Introduction

In polycrystalline aggregates microstructure plays an important role in the evolution of stresses and strains and consequently development of damage processes such as for example evolution of microstructurally small cracks and fatigue. Random grain shapes and sizes, combined with different crystallographic orientations, inclusions, voids and other microstructural features result in locally anisotropic behavior of the microstructure with direct influence on the damage initialization and evolution (Hussain, 1997; Hussain et al., 1993; King et al., 2008a; Miller, 1987). To account for these effects grain-scale or meso-scale models of polycrystalline aggregates are being developed and are increasingly being used.

In this chapter we present some of the most often used approaches to modeling polycrystalline aggregates, starting from more simplistic approaches and up to the most state-of-the art approaches that draw on the as-measured properties of the microstructure. The models are usually based on the finite element approach and differ by a) the level to which they account for the complex geometry of polycrystalline aggregates and b) the sophistication of the used constitutive model. In some approaches two dimensional models are used with grains approximated using simple geometrical shapes like rectangles (Bennett & McDowell, 2003; Potirniche & Daniewicz, 2003) and hexagons (Sauzay, 2007; Shabir et al., 2011). More advanced approaches employ analytical geometrical models like Voronoi tessellation in 2D (Simonovski & Cizelj, 2007; Watanabe et al., 1998) and 3D (Cailletaud et al., 2003; Diard et al., 2005; Kamaya & Itakura, 2009; Simonovski & Cizelj, 2011a). In the most advanced approaches, however, grain geometry is based on experimentally obtained geometry (Lewis & Geltmacher, 2006; Qidwai et al., 2009; Simonovski & Cizelj, 2011b) using methods such as serial sectioning or X-ray diffraction contrast tomography (DCT) (Johnson et al., 2008; Ludwig et al., 2008). These approaches are often referred to as "image-based computational modeling" and can also embed in the model measured properties such as crystallographic orientations. The acquired information is of immense value for advancing our understanding of materials and for developing advanced multiscale computational models. The rather high level of available details may render extremely complex geometries, resulting in highly challenging preparation of finite element (FE) models (Simonovski & Cizelj, 2011a) and computationally extremely demanding simulations. These two constraints have so far limited the development and use of the image-based models. Steps aimed at obtaining a 'reasonable' size model in the terms of computational times are presented. The term 'reasonable' should be taken in relative terms as these models may still run for several days on today's high performance clusters.

Specifically, two analytical approaches (two and three dimensional Voronoi tessellations) for modeling grain geometries as well as an approach based on the X-ray diffraction contrast tomography (DCT) (Johnson et al., 2008; Ludwig et al., 2008) are presented. The DCT enables spatial non-destructive characterization of polycrystalline microstructures (King et al., 2010). The process of building a finite element model from either analytical spatial structures or as-measured spatial structures is explained. The meshing procedure, including ensuring conformal mesh between the grains as well as grain boundary modeling are discussed and explained. The most used constitutive models and the issues related to the modeling of the grain boundaries using the cohesive zone approaches are discussed. In the last section the effects of grain structure to inhomogeneous stress/strain distribution is demonstrated. Initiation and development of intergranular stress corrosion cracks is outlined and discussed for different constitutive models. Also, the stability of the simulations and measures aimed at improving it are considered.

2. Analytical models of structures

In material science Voronoi tessellations are extensively used to model grain geometry with the purpose of calculating the properties of polycrystalline aggregates (Cailletaud et al., 2003; Kovač & Cizelj, 2005), modeling short crack initiation and propagation or modeling intergranular stress corrosion cracks (Kamaya & Itakura, 2009; Musienko & Cailletaud, 2009; Simonovski & Cizelj, 2011b). A Voronoi tessellation is a cell structure constructed from randomly positioned points, also referred to as Poisson points. For polycrystalline aggregates we can think of these points as points where the solidification starts and then uniformly extends in all directions. A solidification front then expands until it collides with another one, thus creating a grain boundary. In geometrical terms the grain boundary is obtained by introducing lines perpendicular to lines connecting neighboring Poisson points. The result is a set of convex polygons/polyhedra, see Fig. 1. Examples of 2D Voronoi tessellations can be seen in Fig. 2. Additional details on mathematical background and applications are available in (Aurenhammer, 1991; Okabe et al., 2000).

Three dimensional Voronoi tessellations can be created using the same approach. A number of mathematical and programming packages like Matlab have the option of constructing such tessellations. Fig. 3 shows examples of 3D Voronoi tessellations generated using Qhull algorithm (*Qhull code for Convex Hull, Delaunay Triangulation, Voronoi Diagram, and Halfspace Intersection about a Point*, n.d.) and implemented in (Petrič, 2010).

3. As-measured spatial structures

Models of polycrystalline aggregates can also be created from the experimental data, where the grain shapes and orientations are measured, i.e. the microstructure is being characterized. A widely used method for characterizing the microstructure in order to get the data for creating a finite element model is serial sectioning (Lewis & Geltmacher, 2006; Qidwai et al., 2009; Spowart et al., 2003). With this approach a surface of the specimen is characterized and then a thin layer of material is removed to be able to characterize the next layer. The procedure is consecutively repeated, thus obtaining the data at different depths. The 3D

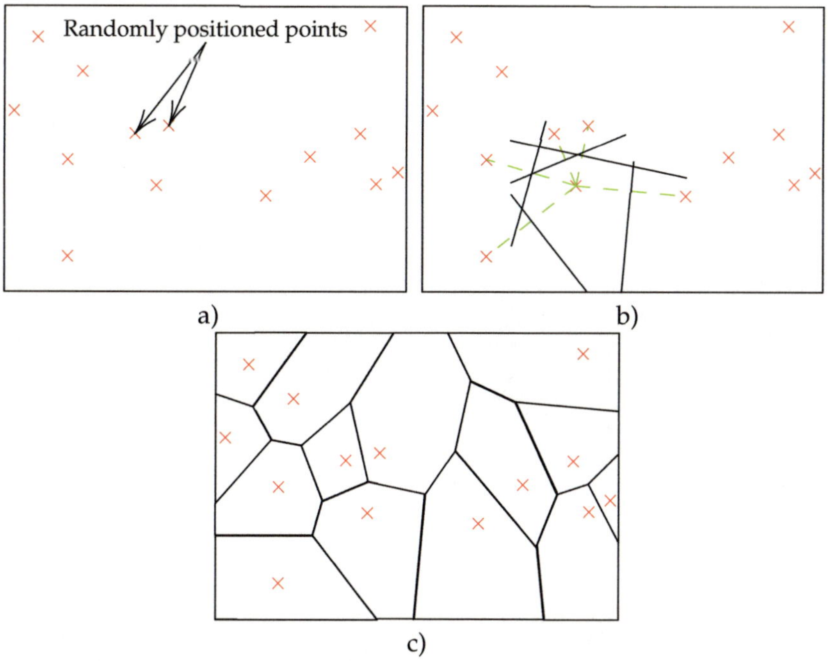

Fig. 1. Construction of a Voronoi tessellation in 2D: a) Poisson points, b) perpendicular lines are introduced to lines connecting neighboring Poisson points and c) final Voronoi tessellation.

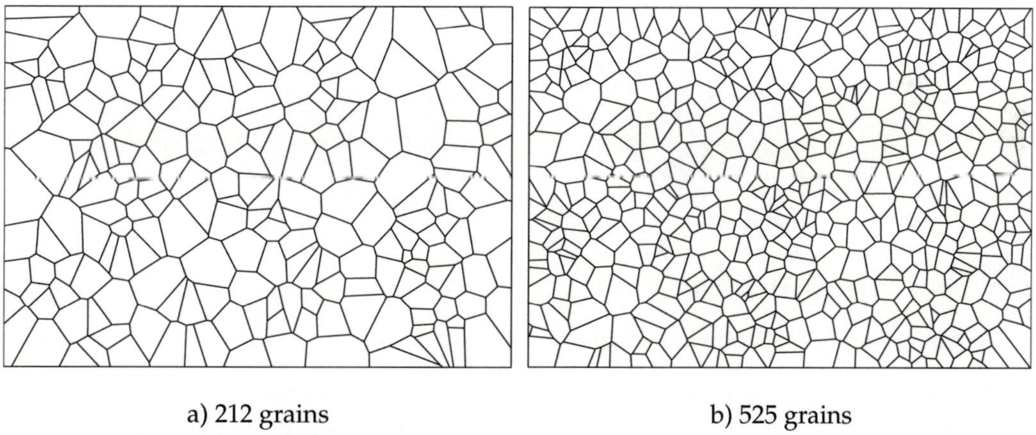

Fig. 2. Examples of 2D Voronoi tessellations.

shapes of the grains are then reconstructed from the data from the 2D layers. However, the problem with this approach is that the specimen is destroyed during the measurement procedure. In last years new experimental techniques have enabled non-destructive spatial characterization of polycrystalline aggregates. Differential aperture X-ray microscopy (Larson et al., 2002), 3D X-ray diffraction microscopy (3DXRD) (Poulsen, 2004) and X-ray diffraction contrast tomography (DCT) (Johnson et al., 2008; Ludwig et al., 2008) are examples of these

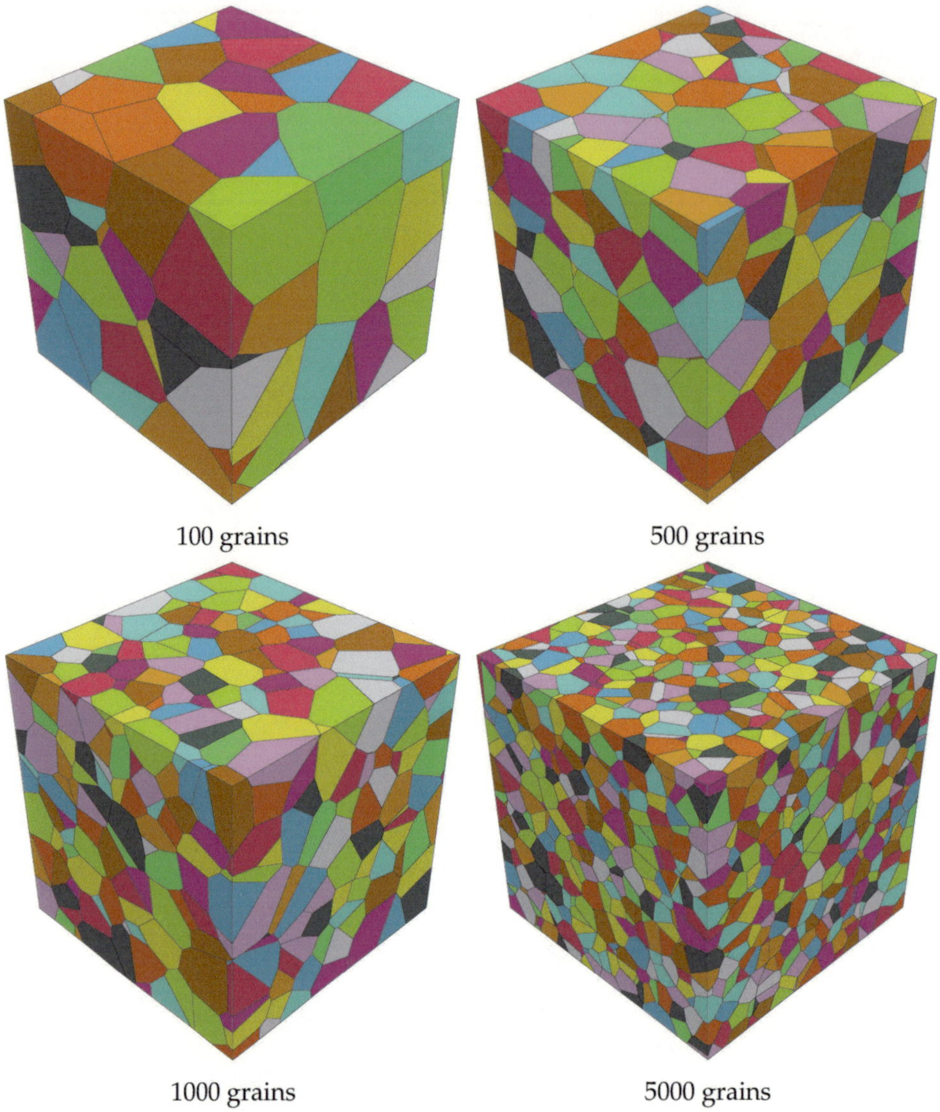

Fig. 3. Examples of 3D Voronoi tessellations.

procedures. Through DCT for example, grain shapes and orientations can be measured and even crack initiation and growth can be monitored (Herbig et al., 2011). Since for some of the presented cases the data has been acquired using DCT, the next section gives an overview of this technique.

3.1 X-ray diffraction contrast tomography (DCT)

DCT (King et al., 2008a; Ludwig et al., 2008) is a measurement procedure jointly developed by the European Synchrotron Radiation Facility (ESRF) and University of Manchester, Materials Performance Centre, School of Materials. It combines the X-ray diffraction imaging and image reconstruction from projections to obtain the data on the grain shapes and crystallographic

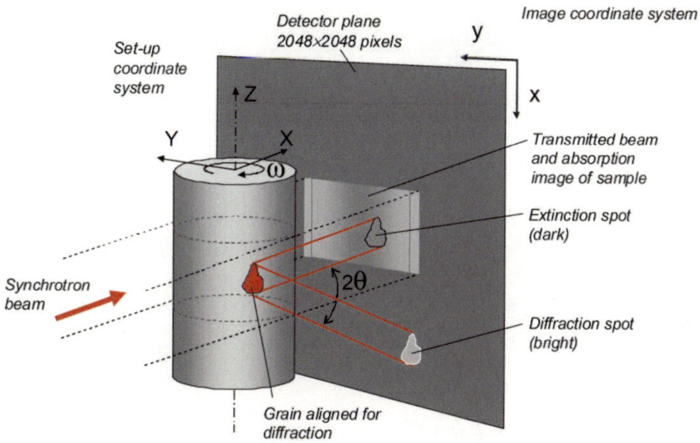

Fig. 4. A scheme of X-ray diffraction contrast tomography (DCT), after (Johnson et al., 2008).

orientations. A rotating polycrystalline sample is exposed to a monochromatic X-ray wave while the projection images are recorded, Fig. 4. Since the sample is rotating, each grain will pass through Bragg diffraction alignments several times. A detector system, significantly bigger than the sample, captures low index reflections. In absence of orientation and strain gradients within the grains, the diffracted beams form 2D spots that can be treated as parallel projections of the grains' volume (King et al., 2010). The shape of each grain can then be reconstructed in 3D using algebraic reconstruction techniques (Gordon et al., 1970). The resolution of the technique is in the order of 1 μm. An example of a measured grain shape of a 400 μm diameter stainless steel wire is given in Fig. 5. The complete wire is depicted in Fig. 6.

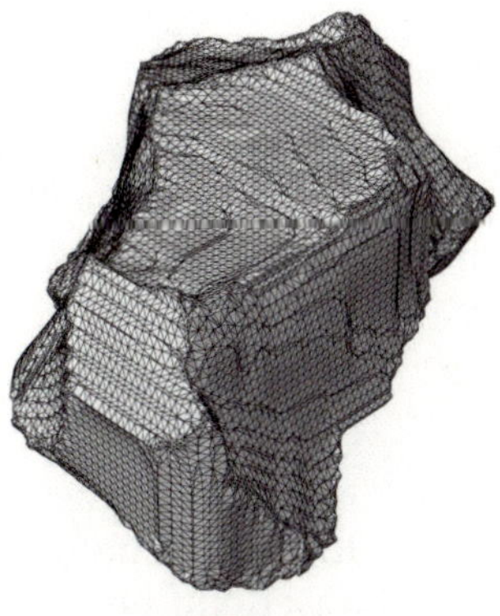

Fig. 5. An example of a measured grain shape. Full resolution is used.

One can see that the available level of detail is very high and that the obtained geometry is extremely complex. This results in highly challenging preparation of finite element (FE) models and computationally extremely demanding simulations. These two constraints have so far limited the development and use of the image-based models. However, with suitable simplifications and parallel pre-processing (Simonovski & Cizelj, 2011a), appropriate FE models can be built in a reasonable time.

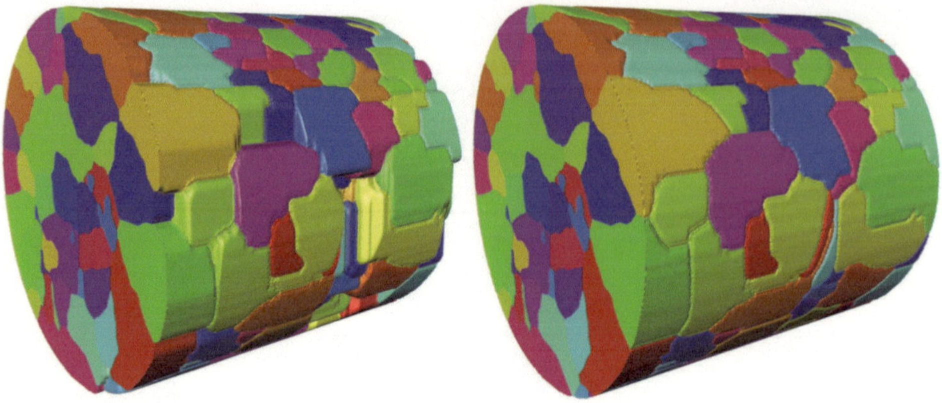

Fig. 6. Reconstructed geometry of a 400 μm diameter stainless steel wire. Left: holes in the original experimental data. Right: wire after the hole treatment, surface grooves present.

3.2 Experimental data

The experimental data used in this work is of a 400 μm diameter stainless steel wire characterized in 3D by DCT (King et al., 2008a). The data has been kindly provided by the University of Manchester, Materials Performance Centre, School of Materials and comprises of 362 grains and some 1600 grain boundaries. The data provides information on the crystallographic orientation in points of a 346 by 346 by 282 grid. The experimental data can be represented as an array of 282 slices, separated in the depth (Z) direction by 1.4 μm. For each slice, crystallographic orientation has been measured on a 346 by 346 grid with 1.4 μm distance between the points on a grid (in the X and Y direction). Voxels having the same crystallographic orientation constitute a grain.

3.3 From the measured data to the surfaces

DCT characterization of a polycrystalline aggregate results in voxel-based data. Voxel-based data is also obtained in other experimental techniques like computed tomography (CT) or magnetic resonance imaging (MRI). To obtain the shapes of individual grains their surfaces need to be reconstructed from the voxel-based data.

3.3.1 Treatment of holes

The original DCT data contains 'holes' in the reconstructed grains, see the left-hand-side of Fig. 6. These are typical artifacts due to the limited number of projections available for each grain and the presence of erroneous contrast (Johnson et al., 2008; Ludwig et al., 2008). Since the holes are not expected in the experimental data, a simple and efficient treatment algorithm can be used to fill the holes by grain growth:

- The layers with depth of one voxel are successively added to the grains in the holes vicinity. Each layer is defined along the border between grains and holes by inspecting the one voxel deep neighborhood of the voxels within holes. Only the hole voxels with exactly one neighboring grain are added to this grain within each layer.
- Only the hole voxels with multiple neighbors remain after the first step. These are assigned to the grains dominating their immediate (one voxel deep) neighborhood.

The right-hand-side of Fig. 6 displays wire data-set after the treatment above.

3.3.2 Surface reconstruction

Geometries of individual grains need to be reconstructed from the voxel-based data. This is usually achieved through reconstructing the surfaces of individual grains. Surface reconstruction from voxels is available in a number of commercial visualization tools. The origins of these tools can mainly be traced to the field of medical visualization. The tools were later further developed for the application to material science. In this work surfaces are reconstructed as sets of triangles with Amira package (Visage Imaging GmbH, 2010). A label is assigned to each measured point, defining to which grain this point belongs. The labels are equal to the index of the crystallographic orientation. Label 1 refers to grain 1 with crystallographic orientation index 1 and so forth.

Amira's built-in SurfaceGen tool with unconstrained smoothing option is used. This tool partitions the bounding volume into 362 grains depending on the number of different labels in the 8 vertices of a given voxel. Near the triple points between the grains and near the grain boundaries vortices of a given voxel will be distributed among several grains. In these cases the voxel is subdivided into up to 6^3 sub-vortexes to give a topologically correct representation of the implicitly defined separating surfaces (Westerhoff, 2003). If two adjacent sub-vortexes are of different grains, their common face is added to the list of boundaries between the two grains. A comprehensive explanation of the procedure is given in (Stalling et al., 1998; Westerhoff, 2003) and was later implemented in Amira. Described approach automatically increases the resolution near the triple points between the grains and near the grain boundaries where vortices of a given voxel are distributed among several labels/grains. This is especially important since stress increases at these points can be expected due to different crystallographic orientations of the adjacent grains. Further details on the implemented approach can be found in (Simonovski & Cizelj, 2011a).

The density of the triangles forming the reconstructed surfaces is limited by the resolution of the experimental data. At full resolution the number of triangles is 4 758 871, resulting in FE model with 51 211 552 finite elements. The number of triangles therefore needs to be decreased. This is done using Amira's built-in surface simplification tool (Zachow et al., 2007). The simplification decreases the details as well as resolution at the triple lines between the grains, see Fig. 7 where the number of triangles has been decreased to 30 000 (case 30K), 150 000 (case 150K) and 300 000 (case 300K).

4. From surfaces to FE models

The complexity of the reconstructed surfaces, together with a rather large number of grains, essentially prevents the direct use of the finite element meshing capabilities of either Amira or even professional mesh generators such as for example ABAQUS/CAE (Simulia, 2010). A framework for the automatic meshing has been therefore developed (Simonovski & Cizelj, 2011a) using the Python scripting language and ABAQUS/CAE meshing tools, which are fully accessible through Python.

The framework can be applied to both analytical models of structures (e.g. 3D Voronoi tessellations) and to data obtained from experimental techniques. In both cases the surface structures are defined by the triangle-based surfaces, bounding the volume of individual grains. In the case of voxel-based data these surfaces are reconstructed with Amira. For

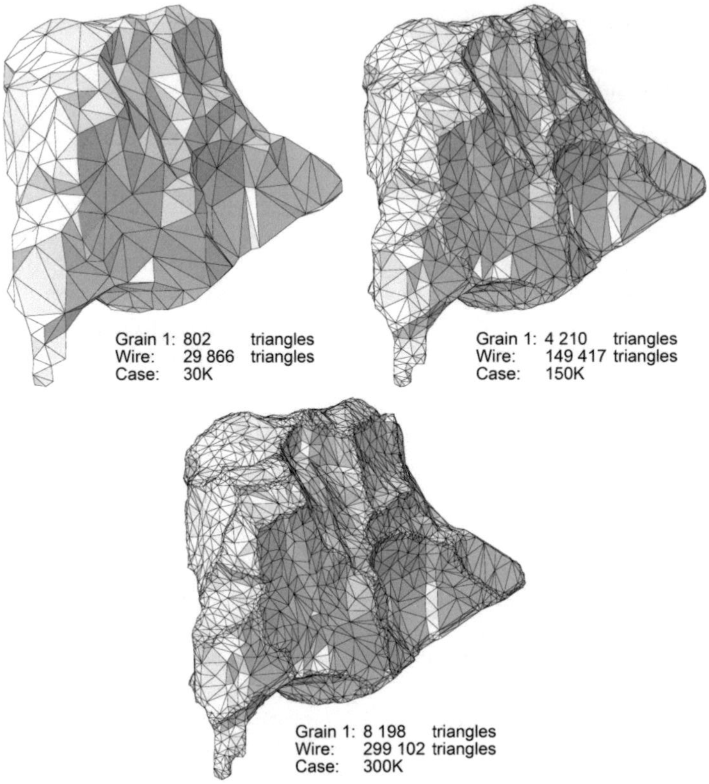

Fig. 7. Geometry reconstructed from experimental data: grain 1 from the wire data set.

analytical models such as 3D Voronoi tessellations, the spatial structure is generated by the underlying analytical model (e.g. Qhull algorithm (*Qhull code for Convex Hull, Delaunay Triangulation, Voronoi Diagram, and Halfspace Intersection about a Point*, n.d.) implemented in (Petrič, 2010)) and surface reconstruction is not needed.

Before starting the FE meshing procedure the surface triangles aspect ratios are checked. Surface triangle aspect ratio is defined in this work as the ratio of the circumscribed circle and the inscribed circle of a triangle. Triangles with aspect ratio of more than 1000 are removed by collapsing triangle's shortest edge, removing the triangle from the structure and updating the vertices and triangles. The procedure is performed iteratively until the worst aspect ratio is above 1000. This approach improves the FE mesh quality. The triangle-based surfaces are also checked for possible errors like intersections and corrected, if necessary, by slight displacement of appropriate triangle's corner points.

The ability of exporting the reconstructed surfaces into a standard CAD format that can be read by FE pre-processors is lacking in many visualization tools. Instead, the user is encouraged to use the tool's built in meshers, which often do not match the capabilities of dedicated FE mesh engines. Furthermore, FE pre-processors do not support export formats of the visualization tools. So there is a basic difficulty of importing the reconstructed geometry into FE pre-processors. This issue is circumvented here by developing a function for exporting the reconstructed surfaces into ACIS SAT file that can be imported into practically any FE pre-processor. The surfaces of each grain are therefore saved into ACIS SAT file with all the surface triangles that they contain. Next, individual ACIS SAT file is imported into ABAQUS/CAE pre-processor, assigned seeds and its surfaces are meshed using triangular FE elements of a selected size. Mesh density is user controlled. The coarsest possible mesh is defined by the triangles of the reconstructed surface. Finer meshes can be obtained by using several FE elements per each triangle of the reconstructed surface. Meshing individual surfaces is independent upon each other which provides for an efficient parallelization of the process. A surface between the adjoining grains needs to be meshed only once since the shared surface is identical for both grains.

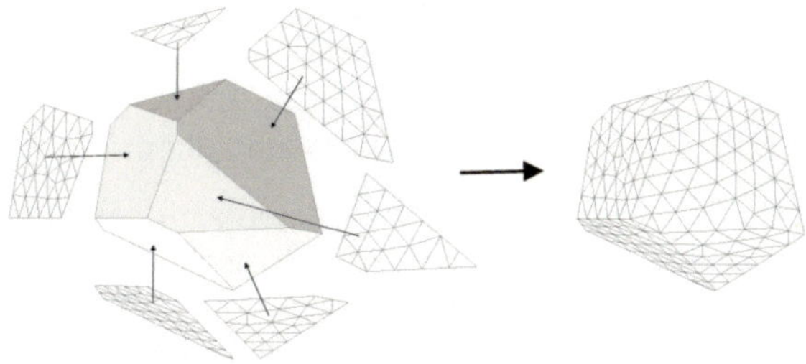

Fig. 8. Creating a conformal volume mesh between grains by meshing the grain boundary surfaces and imprinting the obtained surface meshes on the corresponding grain boundaries.

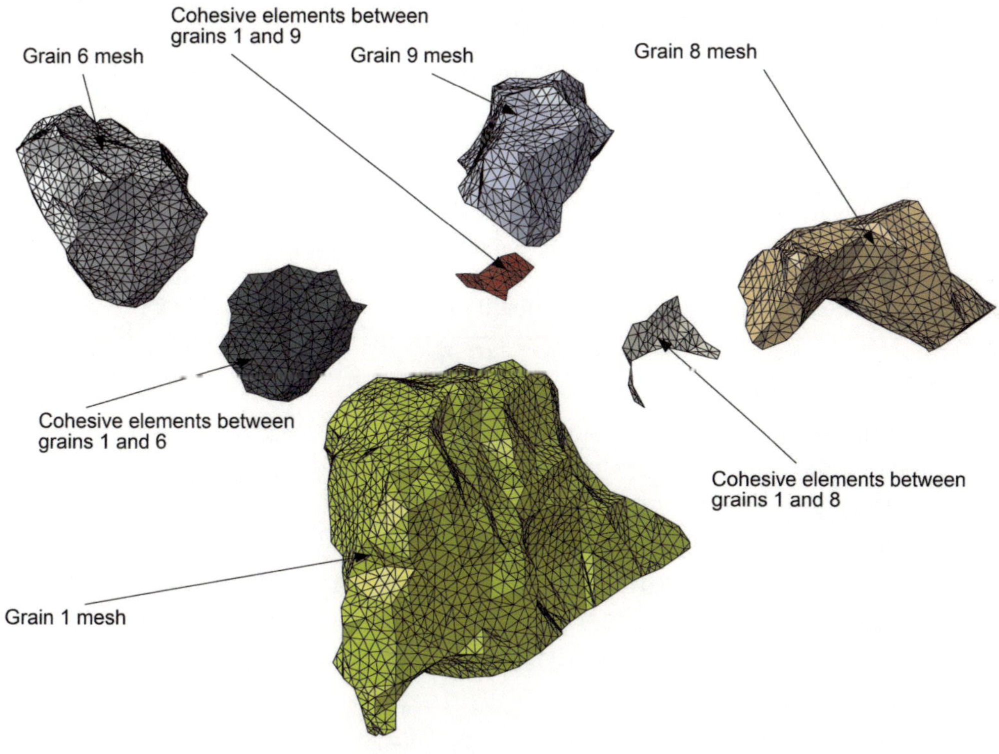

Fig. 9. Detail of the FE mesh showing cohesive elements between the grains.

To obtain a conformal mesh between adjoining grains the meshed surfaces are imprinted onto the corresponding grains, see Fig. 8. This is done by replacing the reconstructed surface of a given grain with an appropriate FE meshed surface and saving the grains' new geometry with all the imprinted meshes into new ACIS SAT files. The process is repeated for all the grains. Updated ACIS SAT geometry files of individual grains with imprinted surface meshes are imported into a ABAQUS/CAE, assigned appropriate surface definitions, material properties, loads and boundary conditions. Exactly one FE seed per each edge is assigned to preserve the FE meshed surfaces, obtained in the previous step. The number of FE per edge is not allowed to increase/decrease thus automatically creating conformal meshes between adjoining constituents. FE volume-meshing is performed next using ABAQUS/CAE built-in mesher. All the information that has been generated in the previous steps (topology, common surfaces between the constituents, material properties,...) has been saved using the Python pickle module and is now used to hierarchically define all the properties, including node, element and surface sets. Generating FE models of individual grains is independent upon each other which again provides for an efficient parallelization.

In the last step, zero thickness layers of cohesive elements are inserted between the adjacent grains. Layers of zero thickness triangular cohesive elements are inserted between the nodes occupying the same position on the adjacent surfaces. The triangular cohesive elements are oriented to conform with the tetrahedral elements on both surfaces. The nodes, elements, set, surfaces,... and all other definitions are also updated to reflect the new configuration. Fig. 9 illustrates mesh of adjacent grains with inserted zero thickness cohesive layers. Further details on the procedures employed in automatic and parallel generation of the finite element meshes are available in (Simonovski & Cizelj, 2011a).

Fig. 10 shows the obtained FE models for the 3D Voronoi tessellations given in Fig. 3. The mesh quality factors are given in Table 1.

Fig. 11 illustrates the constructed FE model of the 400 μm diameter stainless steel wire. The top figure corresponds to the first experimental series, containing 362 grains. The model contains 903 199 finite elements: 796 105 linear solid tetrahedra elements and 107 094 cohesive elements. The bottom figure corresponds to the second experimental series, containing 1334 grains. The model contains 3 395 769 finite elements: 2 976 828 linear solid tetrahedra elements and 418 941 cohesive elements. The mesh quality factors are given in Table 2.

	100 grains	500 grains	1000 grains	5000 grains
Number of elements				
Solid elements	91 140	143 588	528 860	4 517 884
Cohesive elements	13 363	34 136	105 468	787 976
All elements	104 503	177 724	634 328	5 305 860
Number of elements with				
Min angle < 5 [°]	38 (0.0417 %)	1021 (0.7111 %)	1571 (0.2970 %)	7847 (0.1737 %)
Max angle > 170 [°]	0 (0 %)	11 (0.0077 %)	4 (0.0007 %)	27 (0.0006 %)
Aspect ratio > 10	43 (0.0472 %)	1103 (0.7682 %)	1742 (0.3294 %)	8651 (0.1915 %)
Values of				
Worst min angle [°]	2.52	0.02	0.1	0.05
Worst max angle [°]	162.09	179.9	172.5	174.27
Worst aspect ratio	27.48	1156	512.4	1028

Table 1. 3D Voronoi FE models: mesh quality factors comparison.

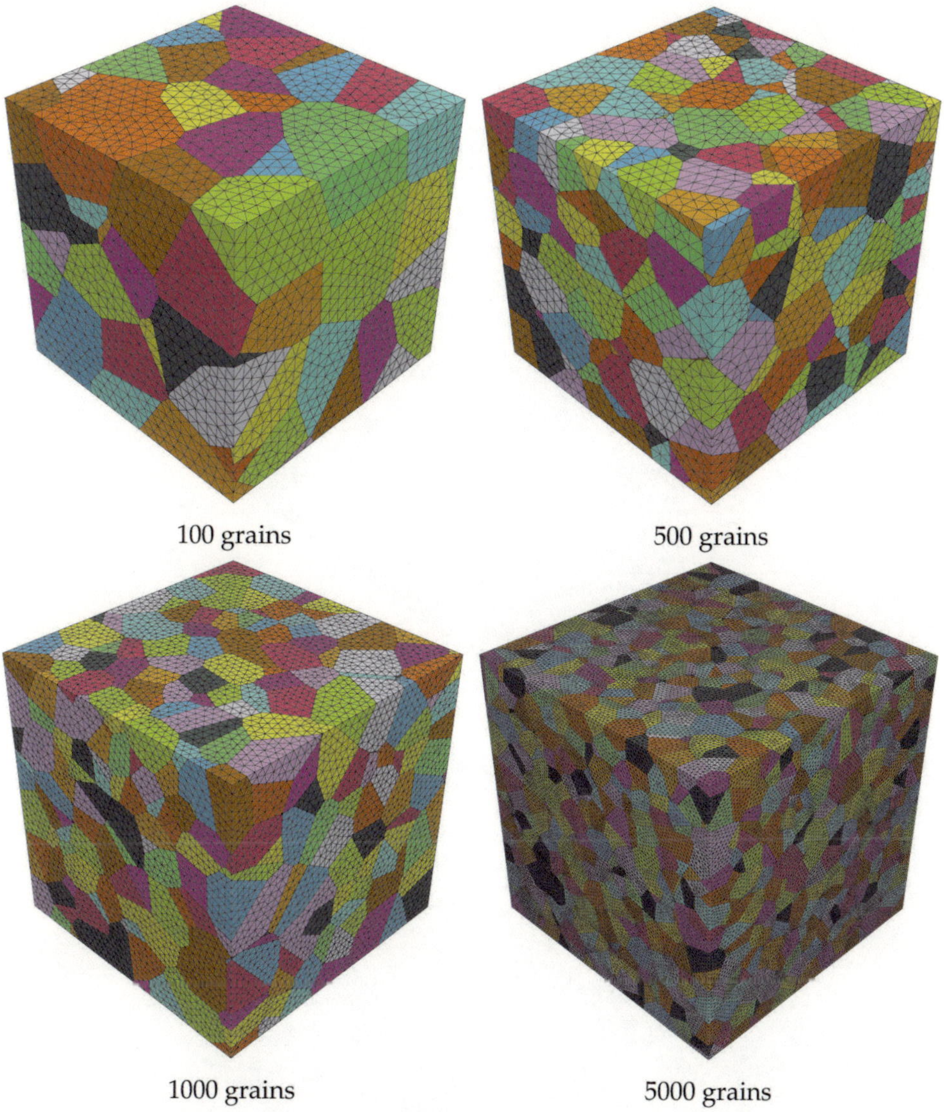

Fig. 10. FE mesh examples of 3D Voronoi tessellations.

5. Constitutive models

5.1 Bulk grains

Isotropic elasticity, anisotropic elasticity and anisotropic elasticity with crystal plasticity constitutive laws are commonly used for bulk grains. Since isotropic elasticity can not account for the effects due to different crystallographic orientations of the grains this is not covered here. Overview of the other two constitutive laws is given below.

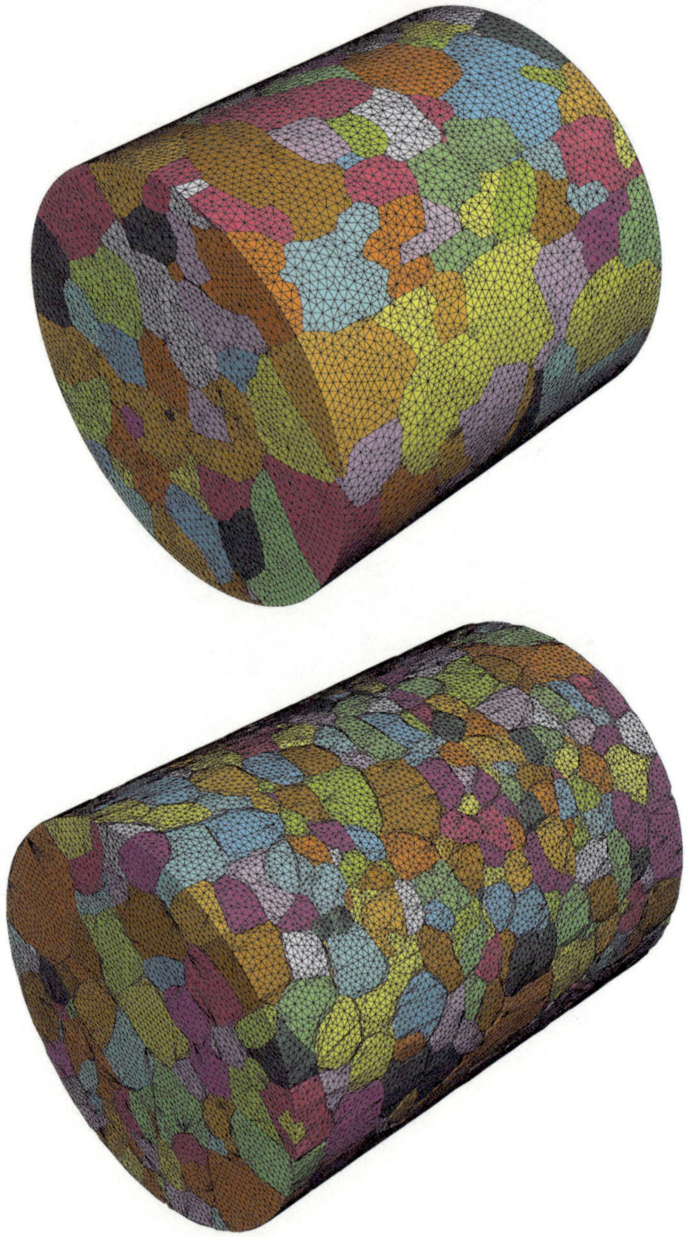

Fig. 11. A FE model of the 400 μm diameter stainless steel wire. Top: wire from the first experimental series, 362 grains. Bottom: wire from the second experimental series, 1334 grains.

	362 grain wire	1334 grain wire
Number of elements		
Solid elements	796 105	2 976 828
Cohesive elements	107 094	418 941
All elements	903 199	3 395 769
Number of elements with		
Min angle < 5 [°]	32298 (0.2886 %)	174 (0.0058 %)
Max angle > 170 [°]	45 (0.0056 %)	12 (0.0004 %)
Aspect ratio > 10	2449 (0.3076 %)	152 (0.0051 %)
Values of		
Worst min angle [°]	0.09	0.47
Worst max angle [°]	178.77	178.64
Worst aspect ratio	647.1	23.42

Table 2. Wire FE models: mesh quality factors comparison.

5.1.1 Anisotropic elasticity

In general, the relation between the elastic stress tensor, $\sigma_{ij}^{El.}$, and the elastic strain tensor, $\epsilon_{kl}^{El.}$ for anisotropic elasticity is given by Eq. (1).

$$\sigma_{ij}^{El.} = C_{ijkl} \cdot \epsilon_{kl}^{El.} \tag{1}$$

Here, the C_{ijkl} stands for the stiffness tensor. For materials with a cubic lattice there are only three independent values in the stiffness tensor: $C_{iiii} = C_{11}$, $C_{iijj} = C_{12}$ and $C_{ijij} = C_{44}$. The relation between the strains and stresses is then given by Eq. (2).

$$\begin{Bmatrix} \sigma_{11} \\ \sigma_{22} \\ \sigma_{33} \\ \sigma_{12} \\ \sigma_{23} \\ \sigma_{31} \end{Bmatrix}^{El.} = \begin{bmatrix} C_{11} & C_{12} & C_{12} & 0 & 0 & 0 \\ C_{12} & C_{11} & C_{12} & 0 & 0 & 0 \\ C_{12} & C_{12} & C_{11} & 0 & 0 & 0 \\ 0 & 0 & 0 & C_{44} & 0 & 0 \\ 0 & 0 & 0 & 0 & C_{44} & 0 \\ 0 & 0 & 0 & 0 & 0 & C_{44} \end{bmatrix} \begin{Bmatrix} \epsilon_{11} \\ \epsilon_{22} \\ \epsilon_{33} \\ 2\epsilon_{12} \\ 2\epsilon_{23} \\ 2\epsilon_{31} \end{Bmatrix}^{El.} \tag{2}$$

5.1.2 Crystal plasticity

Crystal plasticity theory (Hill & Rice, 1972; Rice, 1970) assumes that the plastic deformation in monocrystals takes place via a simple shear on a specific set of slip planes. Deformation by other mechanisms such as for example diffusion, twinning and grain boundary sliding is here not taken into account. The combination of a slip plane, denoted by its normal m_i^{α}, and a slip direction, s_i^{α}, is called a slip system, (α). The plastic velocity gradient, $\dot{u}_{i,j}^{p}$, due to a crystallographic slip can be written using Eq. (3) (Needleman, 2000). The summation is performed over all active slip systems,(α), with $\dot{\gamma}^{(\alpha)}$ representing the shear rate.

$$\dot{u}_{i,j}^{p} = \sum_{\alpha} \dot{\gamma}^{(\alpha)} s_i^{(\alpha)} m_j^{(\alpha)} \tag{3}$$

The cumulative slip is defined as $\gamma = \sum_\alpha \int_0^t \left|\dot{\gamma}^{(\alpha)}\right| dt$. From the well-known relation for small strain $\epsilon_{ij} = \frac{1}{2}\left(u_{i,j} + u_{j,i}\right)$ one can obtain the plastic strain rate, Eq. (4). The constitutive relation of the elastic-plastic monocrystal is now given in terms of stress and strain rates as $\dot{\sigma}_{ij} = C_{ijkl}\left(\dot{\epsilon}_{kl} - \dot{\epsilon}^p_{kl}\right)$ (Needleman, 2000).

$$\dot{\epsilon}^p_{ij} = \sum_\alpha \frac{1}{2}\dot{\gamma}^{(\alpha)}\left(s_i^{(\alpha)} m_j^{(\alpha)} + s_j^{(\alpha)} m_i^{(\alpha)}\right) \tag{4}$$

It is assumed that the shear rate $\dot{\gamma}^{(\alpha)}$ depends on the stress only via the Schmid resolved shear stress, $\tau^{(\alpha)}$, Eq. (5) and Eq. (6). This is a reasonable approximation at room temperature and for ordinary strain rates and pressures (Needleman, 2000). Yielding is then assumed to take place when the Schmid resolved shear stress exceeds the critical shear stress τ_0.

$$\dot{\gamma}^{(\alpha)} = \dot{a}^{(\alpha)}\left(\frac{\tau^{(\alpha)}}{g^{(\alpha)}}\right)\left|\frac{\tau^{(\alpha)}}{g^{(\alpha)}}\right|^{n-1} \tag{5}$$

$$\tau^{(\alpha)} = s_i^{(\alpha)} \sigma_{ij} m_j^{(\alpha)} \tag{6}$$

$\dot{a}^{(\alpha)}$ represents the reference strain rate, n the strain-rate-sensitivity parameter and $g^{(\alpha)}$ the current strain-hardened state of the crystal. In the limit, as n approaches infinity, this power law approaches that of a rate-independent material. The current strain-hardened state $g^{(\alpha)}$ can be derived from Eq. (7), where $h_{\alpha\beta}$ are the slip-hardening moduli defined by the adopted hardening law.

$$\dot{g}^{(\alpha)} = \sum_\beta h_{\alpha\beta} \quad \dot{\gamma}^{(\beta)} \tag{7}$$

$$h_{\alpha\alpha} = \left\{(h_0 - h_s)\,\mathrm{sech}^2\left[\frac{(h_0 - h_s)\,\gamma^{(\alpha)}}{\tau_s - \tau_0}\right] + h_s\right\} G\left(\gamma^{(\beta)}; \beta \neq \alpha\right) \tag{8}$$

$$h_{\alpha\beta} = q h_{\alpha\alpha}, \qquad (\alpha \neq \beta), \tag{9}$$

$$G\left(\gamma^{(\beta)}; \beta \neq \alpha\right) = 1 + \sum_{\beta \neq \alpha} f_{\alpha\beta} \tanh\left(\frac{\gamma^{(\beta)}}{\gamma_0}\right) \tag{10}$$

In this work the Bassani (Bassani & Wu, 1991) hardening law is used with the hardening moduli defined with Eqs. (8, 9, 10). Here h_0 stands for the initial hardening modulus, τ_0 the yield stress (equal to the initial value of the current strength $g^{(\alpha)}(0)$) and τ_s a reference stress where large plastic flow initiates (Huang, 1991). h_s is hardening modulus during easy glide within stage I hardening and q is a hardening factor. The function G is associated with cross hardening where γ_0 is the amount of slip after which the interaction between slip systems reaches the peak strength, and each component $f_{\alpha\beta}$ represents the magnitude of the strength of a particular slip interaction. This model was implemented as a user-subroutine into the finite element code ABAQUS. Further details on the applicable theory and implementation can be found in (Huang, 1991).

5.2 Grain boundaries with cohesive zone approach

Based on the experimental observations, e.g. (Coffman & Sethna, 2008), grain boundaries are modeled with a cohesive-zone approach using cohesive elements. The traction-separation constitutive behaviour with the damage initiation and evolution as implemented in ABAQUS are used in this work. The cohesive elements are inter-surface elements of often negligible thickness, which essentially measure the relative displacements of the surfaces of adjoining continuum elements. The strains ϵ in the cohesive elements are defined using the constitutive thickness of the element T_0 (mostly different from the geometric thickness which is typically close or equal to zero) and the separations of the element nodes δ as compared to their initial unloaded positions, Eq. (11).

$$\begin{Bmatrix} \epsilon_n \\ \epsilon_s \\ \epsilon_t \end{Bmatrix} = \frac{1}{T_0} \begin{Bmatrix} \delta_n \\ \delta_s \\ \delta_t \end{Bmatrix} \tag{11}$$

The indices n, s and t denote the normal and two orthogonal shear directions of the cohesive element. The normal direction always points out of the plane of the cohesive element. The tractions on the cohesive elements are then given by Eq. (12).

$$\begin{Bmatrix} t_n \\ t_s \\ t_t \end{Bmatrix} = \begin{bmatrix} K_{nn} & 0 & 0 \\ 0 & K_{ss} & 0 \\ 0 & 0 & K_{tt} \end{bmatrix} \begin{Bmatrix} \epsilon_n \\ \epsilon_s \\ \epsilon_t \end{Bmatrix} \tag{12}$$

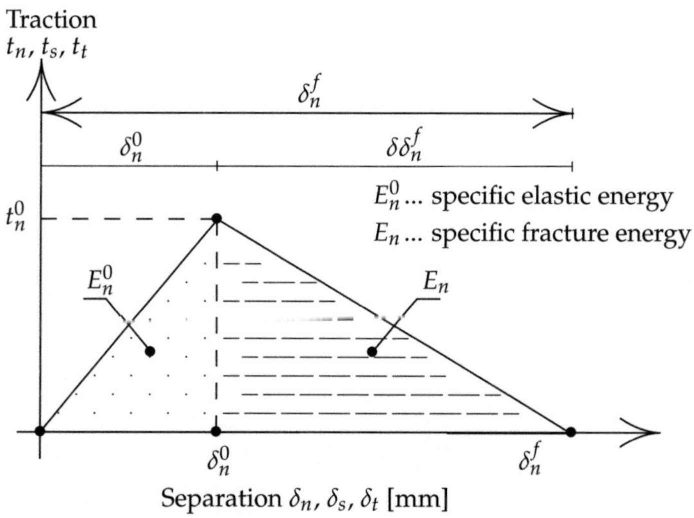

Fig. 12. Example of traction-separation response (not to scale).

Typical traction-separation response is given by Fig. 12. Damage evolution $D(\delta)$ is defined by Eq. (13) for the normal direction (and both shear directions).

$$D(\delta) = \begin{Bmatrix} 0 & ; \delta < \delta_n^0 \\ \frac{\delta_n^f(\delta-\delta_n^0)}{\delta\left(\delta_n^f-\delta_n^0\right)} & ; \delta \geq \delta_n^0 \end{Bmatrix} \tag{13}$$

The actual load-carrying capability of the cohesive element in the normal direction would then be $[1 - D(\delta)]\, K_{nn}$ and correspondingly for the two shear directions.

5.2.1 Cohesive elements issues

In this section we explore the response of the cohesive elements in their normal direction. Let us have a cuboid, divided into three grains as depicted in the Fig. 13 by the three colors. Let us put 100 MPa of tensile stress on the top and the bottom surface and 200 MPa of tensile stress on the left and right surface. Let us constrain the front and the back surface in the Z direction, resulting in ϵ_{33}=0. Furthermore, let us assume that we are dealing with isotropic elastic material with Young modulus E=200 000 MPa and Poisson ratio of ν=0.3. For isotropic elastic material the Eqs. (14,15,16) are valid.

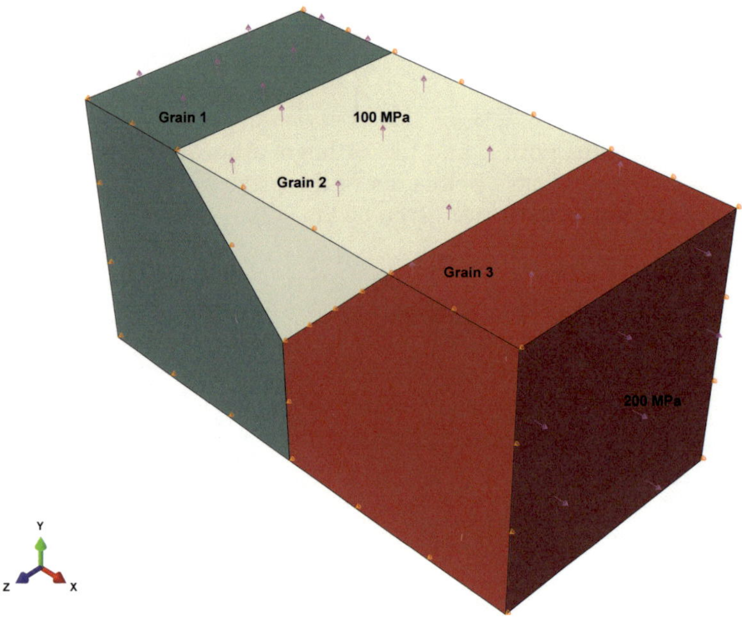

Fig. 13. A simple Y model.

$$\sigma_{11} = \frac{E}{1+\nu}\epsilon_{11} + \frac{E\nu}{(1+\nu)(1-2\nu)}(\epsilon_{11}+\epsilon_{22}+\epsilon_{33}) \tag{14}$$

$$\sigma_{22} = \frac{E}{1+\nu}\epsilon_{22} + \frac{E\nu}{(1+\nu)(1-2\nu)}(\epsilon_{11}+\epsilon_{22}+\epsilon_{33}) \tag{15}$$

$$\sigma_{33} = \frac{E}{1+\nu}\epsilon_{33} + \frac{E\nu}{(1+\nu)(1-2\nu)}(\epsilon_{11}+\epsilon_{22}+\epsilon_{33}) \tag{16}$$

From Eqs. (14,15), expressions Eq. (17,18) can be derived as ϵ_{33}=0 due to the boundary conditions. Using σ_{11}=200 MPa and σ_{22}=100 MPa we obtain values $\epsilon_{22} = 6.5 \cdot 10^{-5}$ and $\epsilon_{11} = 7.15 \cdot 10^{-4}$. Using these two values in Eq. (16) we obtain σ_{33}=90 MPa.

$$\epsilon_{22} = \frac{\sigma_{11}(1+\nu)(1-2\nu) - \sigma_{22}\frac{(1+\nu)(1-2\nu)(1-\nu)}{\nu}}{E\nu - E\frac{(1-\nu)^2}{\nu}} \tag{17}$$

$$\epsilon_{11} = \frac{\sigma_{22}(1+\nu)(1-2\nu) - \epsilon_{22}E(1-\nu)}{E\nu} \tag{18}$$

The resulting stress tensor is given by Eq. (19).

$$\sigma_{ij} = \begin{bmatrix} 200 & 0 & 0 \\ 0 & 100 & 0 \\ 0 & 0 & 90 \end{bmatrix} \text{MPa} \tag{19}$$

The boundaries between the grain are defined with the vectors, normal to the planes of the grain boundaries.

$$n_{Grain1Grain2} = \begin{bmatrix} 1 \\ 1 \\ 0 \end{bmatrix} \cdot \frac{1}{\sqrt{2}} \tag{20}$$

$$n_{Grain2Grain3} = \begin{bmatrix} -1 \\ 1 \\ 0 \end{bmatrix} \cdot \frac{1}{\sqrt{2}} \tag{21}$$

$$n_{Grain1Grain3} = \begin{bmatrix} 1 \\ 0 \\ 0 \end{bmatrix} \cdot \frac{1}{\sqrt{2}} \tag{22}$$

Since we know the stress tensor and the normals for these three planes, we can compute the stresses in the normal direction for each of them. For the plane between the Grain1 and Grain2

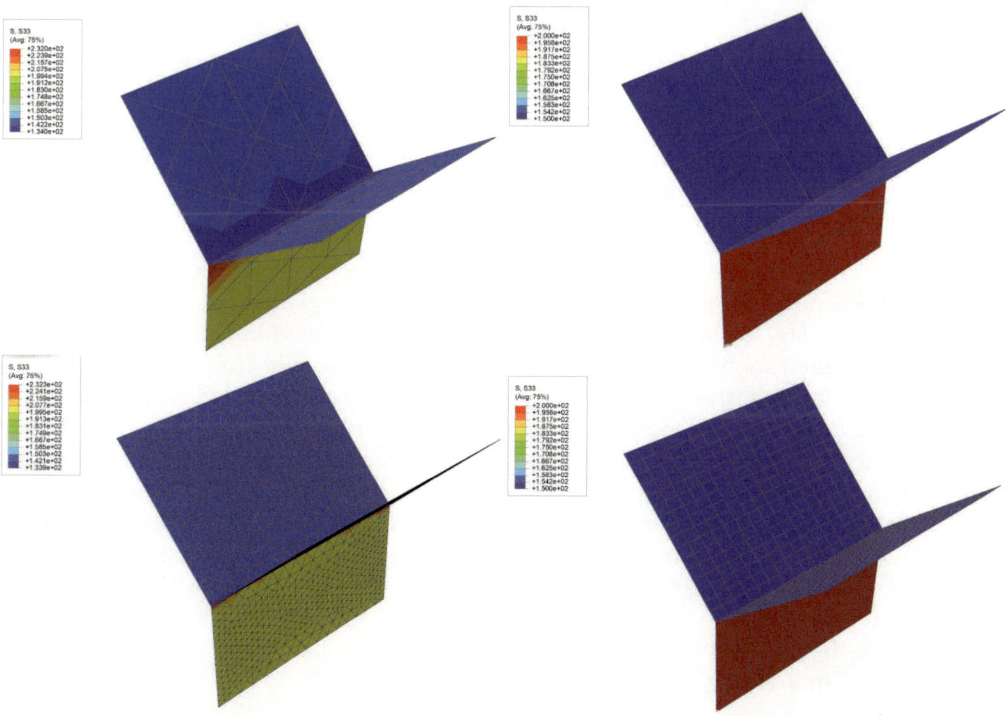

Fig. 14. Normal stresses in the cohesive elements. Triangular prisms (left) and rectangular prisms (right) cohesive elements. Variation in stress at the Y triple points can be observed for the triangular prism cohesive elements.

the stress vector on the plane, p, and the normal stress, σ_n, are given by Eq. (23) and Eq. (24).

$$Grain1Grain2: \quad p = \underbrace{\begin{bmatrix} 200 & 0 & 0 \\ 0 & 100 & 0 \\ 0 & 0 & 90 \end{bmatrix}}_{\sigma_{ij}} \cdot \underbrace{\begin{bmatrix} 1 \\ 1 \\ 0 \end{bmatrix} \cdot \frac{1}{\sqrt{2}}}_{n_{Grain1Grain2}} = \begin{bmatrix} 200 \\ 100 \\ 0 \end{bmatrix} \cdot \frac{1}{\sqrt{2}} \,\text{MPa} \tag{23}$$

$$Grain1Grain2: \quad \sigma_n = p_i \cdot n_i = 150\,\text{MPa} \tag{24}$$

Similarly, we obtain the following normal stress values for the other two planes:

$$Grain2Grain3: \quad \sigma_n = 150\,\text{MPa} \tag{25}$$

$$Grain1Grain3: \quad \sigma_n = 200\,\text{MPa} \tag{26}$$

Fig. 14 displays the normal stresses in the cohesive elements as calculated from the ABAQUS finite element models. On the left hand side triangular prism cohesive elements are used, whereas on the right hand side rectangular prism cohesive elements are used. One can see that in the case of triangular prisms there is a variation in the normal stress for a given plane, in particular at the Y triple points. For the rectangular prism cohesive elements no such variations are observed and the values from the FE model match exactly with the theoretically computed values. Rectangular prism cohesive elements should therefore be preferentially used.

Unfortunately, meshing complicated shapes with rectangular prism cohesive elements also requires that the grains need to be meshed with rectangular prisms-hexahedral elements. With shapes as complicated as seen in Fig. 5 this is extremely difficult and one therefore uses triangular prisms-tetrahedral elements. This then requires the use of triangular prism cohesive elements if one is to obtain a conformal mesh between the structural and cohesive elements. One therefore automatically introduces a degree of discrepancy.

Fig. 15 displays the computed versus the theoretical normal stresses for the cohesive elements for the 3D Voronoi tessellation with 100 grains with external load of 70 MPa. Isotropic elastic constitutive law is used for the grains with the wire loaded in tension. The solid line represents the ideal response. One can observe a significant scatter from the ideal response. Fig. 16 displays the cohesive elements with red color indicating elements with more than 50 % difference between the theoretical and computed normal stresses. One can see that a significant scatter exists. Similar scatter has been reported on the grain structure simulated using 3D Voronoi tessellations (Kamaya & Itakura, 2009). Vast majority of the problematic cohesive elements are located on the triple lines between the grains. The scatter of the normal stresses of the cohesive elements not lying on the triple lines is significantly smaller, with most values within the $\pm 20\,\%$ deviation. Increasing mesh density helps to alleviate the issue to some degree by reducing the area of the problematic elements. Other factors such as the stiffness of the cohesive element and its thickness have negligible effect on the scatter (Simonovski & Cizelj, 2011c).

6. Examples

In this section we look at the some results for the wire model with initialization and propagation of intergranular stress corrosion cracks. The grain boundaries are modeled using the above described cohesive zone approach and classified into resistant and susceptible grain

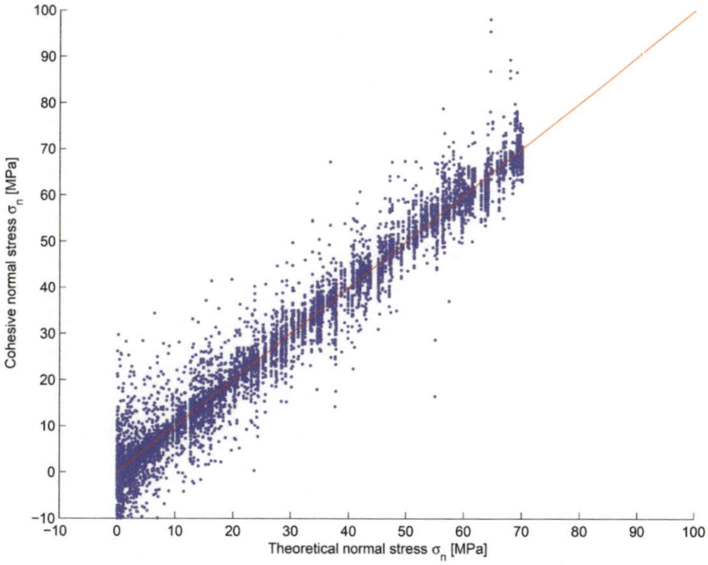

Fig. 15. Theoretical and computed normal stresses at integration points of the cohesive elements. 3D Voronoi, 100 grains, element size=0.025.

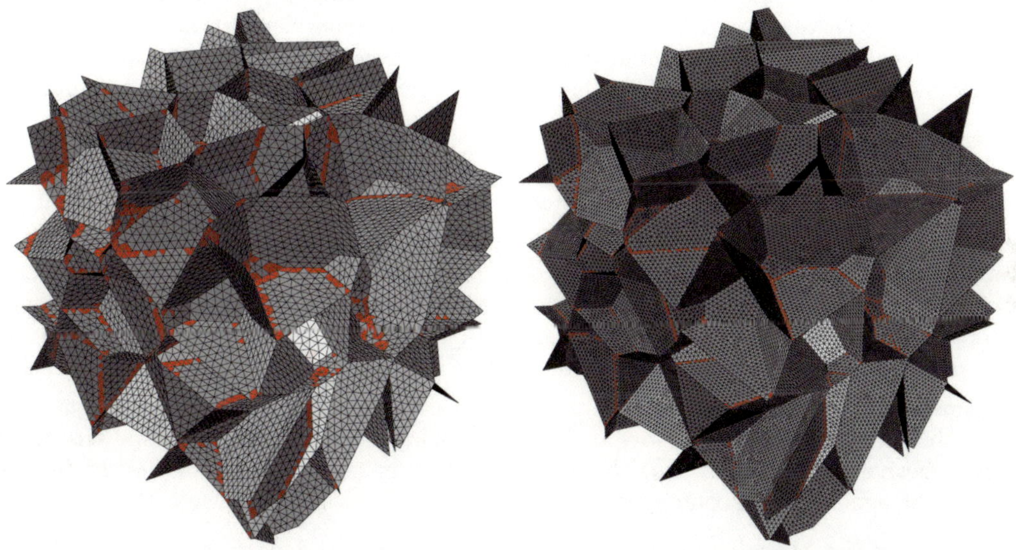

Fig. 16. Cohesive elements with more than 50 % difference between the theoretical and computed normal stresses (in red). 3D Voronoi, 100 grains. Left: element size=0.025. Right: element size=0.0125.

boundaries, depending upon the crystallographic orientation of the neighboring grains. In this work a simplification is used where resistant grain boundaries are defined as coincidence site lattice (Grimmer et al., 1974) ($\Sigma 3$ through $\Sigma 29$) grain boundaries and low angle grain boundaries with misorientation angle between the neighboring grains below 15°, following (Marrow et al., 2006). Σ value is computed as the ratio enclosed by a unit cell of the coincidence sites and the standard unit cell (Bollman, 1982) from the Rodrigues vectors of the two neighboring grains. Brandon (Brandon, 1996) criterion for a proximity to a coincidence site lattice structure with proportionality constant of 10° is used. All other grain boundaries are defined as susceptible grain boundaries.

First, we constrain the nodes on the back surface in all three directions. Next, we apply tensile stress of 60 MPa to the wire's front surface. This load is equal to the one in the experiment (King et al., 2008a;b). After the application of the load we assume that the wire is exposed to an acidified solution of potassium tetrathionate ($K_2S_4O_6$) that penetrates into the wire and degrades the susceptible grain boundaries. This is done to mimic the experiment and is accomplished by employing a user-defined field variable of a cylindrical shape with it's axis aligned with the wire's axis.

The radius of the user-defined field variable cylinder is progressively decreased. Once a cohesive element of a susceptible grain boundary lies outside the user-defined field variable cylinder's radius, its δ_n^0 and $\delta\delta_n^f$ values are decreased, resulting in practically instantaneous full degradation.

Damage of a cohesive element of a susceptible grain boundary that is inside the user-defined field variable cylinder's radius is caused only by mechanical overload, as depicted in Fig. 12. The described approach is purely mechanical. To improve convergence, the decrease of δ_n^0 and $\delta\delta_n^f$ values is not done abruptly. Additionally, a viscous damping of 0.01 is applied to the damage function to improve the convergence during the cohesive elements damage evolution. Susceptible grain boundaries at initial and end 10 % of the wire's length are not allowed to be affected by the user-variable to reduce the edge effect and improve the numerical stability.

The rate at which the radius of the user-defined field variable decreases (i.e. degradation rate) is linked to the stability of the computation. If a large value is selected, then within one computational time increment the separation of the opposite faces in a cohesive element can reach the critical value of δ_n^f at which the element completely degrades. If this degradation process occurs within one computational time increment, the convergence is degraded, even more so when this occurs simultaneously in several cohesive elements. Small enough degradation rates therefore need to be used, so that the process of degradation of cohesive elements is captured within the computational time increments. Alternatively, one can select very small computational time increments or increase the step time. Similarly, the δ_n^0 and $\delta\delta_n^f$ values should not be very small since at already small load increment the resulting separation of the cohesive element faces could be high enough to instantaneously completely damage the element, again causing convergence issues.

Fig. 17 shows the Mises stresses in bulk grains (left part) and damage evolution on the grain boundaries (right part) for the three constitutive models for grains. 30K case with 10.0 μm element size is presented. The first two rows (with the exception of the legend) display the results at an early time increment where the damage due to the corrosion is very limited. For the Mises stress significant differences can be observed between the IE (isotropic elasticity) and AE (anisotropic elasticity) constitutive laws. This is to be expected since the crystallographic orientations are disregarded in IE approach. Due to the low applied load (less than 1/3rd of the yield stress) almost no difference can be observed between the Mises stresses for the AE and AE+CP (anisotropic elasticity+crystal plasticity) constitutive models. This was true even at the end of the simulation where the damage of the grain

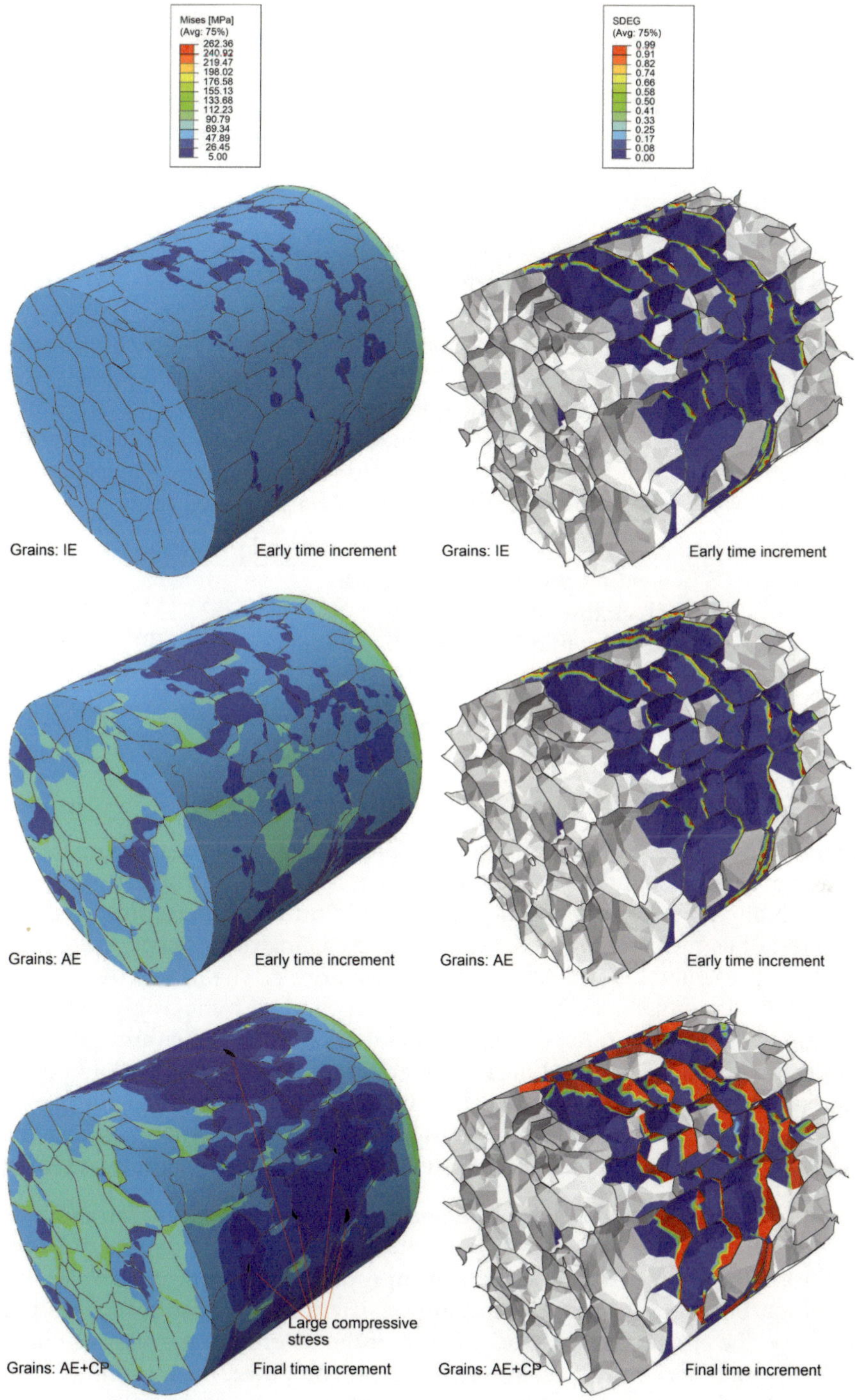

Fig. 17. Mises stress (left) and damage (right) development. Isotropic elastic (top), anisotropic elastic (middle) and anisotropic elastic + crystal plasticity model (bottom).

boundaries was at its highest, resulting in stress redistribution from the failed areas of grain boundaries to the neighboring areas of grains. The last row therefore displays only the Mises stress for the AE+CP model. Redestribution of the Mises stress can be observed due to the degradation of the susceptible, tensile-loaded grain boundaries which decreases the amount of stress transferred from one side of the boundary to the other. Stress is redistributed in the neighboring areas of grain boundaries and grains, increasing the compressive loading. This can be seen as dark patches in the last row.

The presented approach is, however, not without its deficiencies. First, all tensile-loaded grain boundaries degrade at the same rate. In reality, higher degradation rates might be expected for the susceptible grain boundaries with higher tensile load. Also, the initial grain boundary stiffness is taken to be uniform whereas stiffness distribution based upon the properties of adjacent grains is expected (Coffman & Sethna, 2008). Lastly, the selected corrosive environment penetration approach does not account for the topology of the grain boundary network. These issues will therefore need to be addressed in future work. Slightly better convergence of the AE+CP model was observed. However, the computational times for the AE+CP model were more than twice those for the AE model, see Table 3.

	AE	AE+crystal plasticity
30K case, 60 processor cores used		
Wallclock time [s]	252 005	531 133
Memory for min I/O	≈40 GB	≈80 GB
Number of elements	903 199	903 199

Table 3. Model performance data. 10.0 μm element size.

7. Conclusion

With the rapid development of computational capabilities and new experimental techniques we are moving closer to understanding the full role of microstructure on the materials performance. Not only are advanced approaches for simulating microstructures being used but also models of as-measured structures are actively being developed. Among the former, tools such as the presented 2D and 3D Voronoi tessellations can be used whereas for the latter, experimental techniques such as the X-ray diffraction contrast tomography which enable 3D characterization of grains are indispensable. Basics of these approaches are covered here. Both approaches share a common difficult task of creating a finite element model in terms of both appropriate meshes and model sizes. Surface reconstruction issues and complex geometry make the process more difficult. The presented approach effectively deals with some of these issues. The others, however, remain and are subject to further work and research.

The demonstration of the approach is presented on several cases of 3D Voronoi tessellations and two cases of a 400 μm diameter stainless steel wire. In all cases the grain boundaries are explicitly modeled using the cohesive zone approach with zero physical thickness finite elements. Grain boundaries are classified into resistant and susceptible grain boundaries, depending upon the crystallographic orientation of the neighboring grains. Grain boundary damage initialization and early development is then computed for a stainless steel case for several constitutive laws, ranging from isotropic elasticity up to crystal plasticity for the bulk grain material. Since isotropic elasticity approach disregards the crystallographic orientations it should not be used in these cases. Little differences were observed between the anisotropic elasticity and anisotropic elasticity+crystal plasticity approaches, with the latter resulting in more than twice as long computation times.

In all cases almost uniform degradation of the grain boundaries is observed. This is attributed to a) a missing link between the grain boundary load and rate at which the corrosion penetrates the grain boundaries, b) uniform grain boundary stiffnesses whereas stiffness distribution based upon the properties of adjacent grains is expected (Coffman & Sethna, 2008) and c) the selected corrosive environment penetration approach does not account for the topology of the grain boundary network. These issues are the subject of further work.

The numerical stability of the simulation including damage is reasonable, with slightly better convergence for the anisotropic elasticity+crystal plasticity approach. The degradation of a cohesive element is linked to the stability of a simulation. If this degradation process occurs within one computational time increment, the convergence is degraded, even more so when this occurs simultaneously in several cohesive elements. Small computational time increments should therefore be used. Similarly, the δ_n^0 and $\delta\delta_n^f$ values should not be very small since at already small load increment the resulting separation of the cohesive element faces could be high enough to instantaneously completely damage the element, again causing convergence issues.

8. Acknowledgments

The authors would like to thank Prof. James Marrow, (formerly of the University of Manchester, now University of Oxford) and Dr. Andrew King from the European Synchrotron Radiation Facility for kindly providing the results of DCT measurements and also useful discussions. The authors also gratefully acknowledge the financial support from the Slovenian research agency through research programmes P2-0026 and J2-9168.

9. References

Aurenhammer, F. (1991). Voronoi diagrams-a survey of a fundamental geometric data structure, *ACM Computing Surveys* 23(3): 345–405.

Bassani, J. L. & Wu, T.-Y. (1991). Latent hardening in single crystals. II. Analytical characterization and predictions, *Proceedings: Mathematical and Physical Sciences* 435(1893): 21–41.

Bennett, V. P. & McDowell, D. L. (2003). Crack tip displacements of microstructurally small surface cracks in single phase ductile polycrystals, *Engineering Fracture Mechanics* 70(2): 185–207.

Bollman, W. (1982). *Crystal Lattices, Interfaces, Matrices*, Polycrystal Book Service, ISBN 2-88105-000-X.

Brandon, D. (1996). The structure of high-angle grain boundaries, *Acta Metallurgica* 14(11): 1479–1484.

Cailletaud, G., Forest, S., Jeulin, D., Feyel, F., Galliet, I., Mounoury, V. & Quilici, S. (2003). Some elements of microstructural mechanics, *Computational Materials Science* 27(3): 351–374.

Coffman, V. & Sethna, J. (2008). Grain boundary energies and cohesive strength as a function of geometry (http://link.aps.org/doi/10.1103/physrevb.77.144111), *Physical review B* 77(14): 1–11.

Diard, O., Leclercq, S., Rousselier, G. & Cailletaud, G. (2005). Evaluation of finite element based analysis of 3D multicrystalline aggregates plasticity: Application to crystal plasticity model identification and the study of stress and strain fields near grain boundaries, *International Journal of Plasticity* 21(4): 691–722.

Gordon, R., Bender, R. & Herman, G. (1970). Algebraic Reconstruction Techniques (ART) for three-dimensional electron microscopy and X-ray photography, *Journal of Theoretical Biology* 29(3): 471–481.

Grimmer, H., Bollmann, W. & Warrington, D. H. (1974). Coincidence-site lattices and complete pattern-shift in cubic crystals (http://dx.doi.org/10.1107/s056773947400043x), *Acta Crystallographica Section A* 30(2): 197–207.

Herbig, M., King, A.and Reischig, P., Proudhon, H., Lauridsen, E. M., Marrow, J., Buffière, J.-Y. & Ludwig, W. (2011). 3-D growth of a short fatigue crack within a polycrystalline microstructure studied using combined diffraction and phase-contrast X-ray tomography, *Acta Materialia* 59(2): 590–601.

Hill, R. & Rice, J. R. (1972). Constitutive analysis of elastic-plastic crystals at arbitrary strain, *Journal of the Mechanics and Physics of Solids* 20(6): 401–413.

Huang, Y. (1991). A user-material subroutine incorporating single crystal plasticity in the ABAQUS finite element program, *Technical report*, Division of Applied Sciences, Harvard University (http://www.columbia.edu/~jk2079/fem/umat_documentation.pdf).

Hussain, K. (1997). Short fatigue crack behaviour and analytical models: a review, *Engineering Fracture Mechanics* 58(4): 327–354.

Hussain, K., de los Rios, E. & Navarro, A. (1993). A two-stage micromechanics model for short fatigue cracks, *Engineering Fracture Mechanics* 44(3): 425–436.

Johnson, G., King, A., Honnicke, M. G., Marrow, J. & Ludwig, W. (2008). X-ray diffraction contrast tomography: a novel technique for three-dimensional grain mapping of polycrystals. II. The combined case (http://dx.doi.org/10.1107/s0021889808001726), *Journal of Applied Crystallography* 41(2): 310–318.

Kamaya, M. & Itakura, M. (2009). Simulation for intergranular stress corrosion cracking based on a three-dimensional polycrystalline model, *Engineering Fracture Mechanics* 76(3): 386–401.

King, A., Herbig, M., Ludwig, M., Reischig, P., Lauridsen, E., Marrow, T. & Buffière, J. (2010). Non-destructive analysis of micro texture and grain boundary character from X-ray diffraction contrast tomography, *Nuclear Instruments and Methods in Physics Research Section B: Beam Interactions with Materials and Atoms* 268(3-4): 291–296.

King, A., Johnson, G., Engelberg, D., Ludwig, W. & Marrow, J. (2008a). Observations of Intergranular Stress Corrosion Cracking in a Grain-Mapped Polycrystal (http://dx.doi.org/10.1126/science.1156211, http://www.sciencemag.org/cgi/content/full/321/5887/382), *Science* 321(5887): 382 – 385.

King, A., Johnson, G., Engelberg, D., Ludwig, W. & Marrow, J. (2008b). Observations of Intergranular Stress Corrosion Cracking in a Grain-Mapped Polycrystal (http://dx.doi.org/10.1126/science.1156211, http://www.sciencemag.org/cgi/content/full/321/5887/382), *Science* 321(5887): 382 – 385.

Kovač, M. & Cizelj, L. (2005). Modeling elasto-plastic behavior of polycrystalline grain structure of steels at mesoscopic level, *Nuclear Engineering and Design* 235(17-19): 1939–1950.

Larson, B., Yang, W., Ice, G., Budai, J. & Tischler, J. (2002). Three-dimensional X-ray structural microscopy with submicrometre resolution (http://dx.doi.org/10.1038/415887a), *Nature* 415: 887–890.

Lewis, A. & Geltmacher, A. (2006). Image-based modeling of the response of experimental 3d microstructures to mechanical loading, *Scripta Materialia* 55(1): 81–85.

Ludwig, W., Schmidt, S., Lauridsen, E. M. & Poulsen, H. F. (2008). X-ray diffraction contrast tomography: a novel technique for three-dimensional grain mapping of polycrystals. I. Direct beam case (http://dx.doi.org/10.1107/s0021889808001684), *Journal of Applied Crystallography* 41(2): 302–309.

Marrow, T., Babout, L., Jivkov, A., Wood, P., Engelberg, D., Stevens, N., Withers, P. & Newman, R. (2006). Three dimensional observations and modelling of intergranular stress corrosion cracking in austenitic stainless steel, *Journal of Nuclear Materials* 352(1-3): 62–74.

Miller, K. J. (1987). The behaviour of short fatigue cracks and their initiation. Part II-A general summary, *Fatigue & Fracture of Engineering Materials & Structures* 10(2): 93–113.

Musienko, A. & Cailletaud, G. (2009). Simulation of inter- and transgranular crack propagation in polycrystalline aggregates due to stress corrosion cracking, *Acta Materialia* 57(13): 3840–3855.

Needleman, A. (2000). Computational mechanics at the mesoscale, *Acta Materialia* 48(1): 105–124.

Okabe, A., Boots, B., Sugihara, K., Chiu, S. & Chiu, S. (2000). *Spatial Tessellations: Concepts and Applications of Voronoi Diagrams*, 2 edn, Wiley.

Petrič, Z. (2010). Generating 3D Voronoi tessellations for material modeling, *Technical report*, Jozef Stefan Institute.

Potirniche, G. P. & Daniewicz, S. R. (2003). Analysis of crack tip plasticity for microstructurally small cracks using crystal plasticity theory, *Engineering Fracture Mechanics* 70(13): 1623–1643.

Poulsen, H. F. (2004). *Three-Dimensional X-Ray Diffraction Microscopy: Mapping Polycrystals and their Dynamics*, 1st edn, Springer.

Qhull code for Convex Hull, Delaunay Triangulation, Voronoi Diagram, and Halfspace Intersection about a Point (n.d.). http://www.qhull.org.

Qidwai, M., Lewis, A. & Geltmacher, A. (2009). Using image-based computational modeling to study microstructure-yield correlations in metals, *Acta Materialia* 57(14): 4233–4247.

Rice, J. R. (1970). On the structure of stress-strain relations of time-dependent plastic deformation in metals, *Journal of Applied Mechanics* 37: 728–737.

Sauzay, M. (2007). Cubic elasticity and stress distribution at the free surface of polycrystals, *Acta Materialia* 55(4): 1193–1202.

Shabir, Z., Giessen, E., Duarte, C. & Simone, A. (2011). The role of cohesive properties on intergranular crack propagation in brittle polycrystals, *Modelling and Simulation in Materials Science and Engineering* 19(3).

Simonovski, I. & Cizelj, L. (2007). The influence of grains' crystallographic orientations on advancing short crack, *International Journal of Fatigue* 29(9-11): 2005–2014.

Simonovski, I. & Cizelj, L. (2011a). Automatic parallel generation of finite element meshes for complex spatial structures, *Computational Material Science* 50(5): 1606–1618.

Simonovski, I. & Cizelj, L. (2011b). Computational multiscale modeling of intergranular cracking (http://dx.doi.org/10.1016/j.jnucmat.2011.03.051), *Journal of Nuclear Materials* 44(22): 243–250.

Simonovski, I. & Cizelj, L. (2011c). Towards Modeling Intergranular Stress Corrosion Cracks on Grain Size Scales (submitted for publication), *Nuclear Engineering and Design* .

Simulia (2010). *ABAQUS 6.10-1 (http://www.simulia.com/)*.

Spowart, J., Mullens, H. & Puchala, B. (2003). Collecting and Analyzing Microstructures in Three Dimensions: A Fully Automated Approach, *Journal of the Minerals, Metals and Materials Society* 55(10): 35–37.

Stalling, D., Zöckler, M., Sander, O. & Hege, H.-C. (1998). Weighted labels for 3D image segmentation (http://opus.kobv.de/zib/volltexte/1998/383/), *Technical report*, Konrad-Zuse-Zentrum für Informationstechnik Berlin.

Visage Imaging GmbH (2010). *Amira 5.2.1 (http://www.amiravis.com/)*.

Watanabe, O., Zbib, H. M. & Takenouchi, E. (1998). Crystal plasticity: micro-shear banding in polycrystals using voronoi tessellation, *International Journal of Plasticity* 14(8): 771–788.

Westerhoff, M. (2003). *Efficient Visualization and Reconstruction of 3D Geometric Models from Neuro-Biological Confocal Microscope Scans (http://www.diss.fu-berlin.de/diss /receive/ FUDISS_thesis_000000001196)*, PhD thesis, Freie Universitat Berlin, FB Mathematik und Informatik.

Zachow, S., Zilske, M. & Hege, H.-C. (2007). 3D reconstruction of individual anatomy from medical image data: Segmentation and geometry processing (http://opus. kobv.de/zib/volltexte/2007/1072/), *Technical report*, Konrad- Zuse-Zentrum für Informationstechnik Berlin.

Part 2

Methods of Synthesis, Structural Properties Characterization and Applications of Some Polycrystalline Materials

4

NASICON Materials: Structure and Electrical Properties

Lakshmi Vijayan and G. Govindaraj
Department of Physics, School of Physical, Chemical and Applied Sciences, Pondicherry University, R. V. Nagar, Kalapet
India

1. Introduction

Solid electrolytes are one of the functional materials, practically applied in industries because of its high ion conducting property. It provides scientific support for wide variety of advanced electrochemical devices such as fuel cells, batteries, gas separation membranes, chemical sensors and in the last few years, ionic switches. NASICON type ion conductors have been tested widely in energy applications for instance in electric vehicles. High ion conductivity and stability of phosphate units are advantages of NASICON over other electrolyte materials (Hong, 1976). Among the batteries those based on lithium show the best performance.

In NASICON frame-work, $A_xB_y(PO_4)_3$, A is an alkali metal ion and B is a multivalent metal ion. The charge compensating A cations occupy two types of sites, M1 and M2 (1:3 multiplicity), in the interconnected channels formed by corner sharing PO_4 tetrahedra and BO_6 octahedra. M1 sites are surrounded by six oxygens and located at an inversion center and M2 sites are symmetrically distributed around three-fold axis of the structure with ten-fold oxygen coordination. In three-dimensional frame-work of NASICON, numerous ionic substitutions are allowed at various lattice sites. Generally, NASICON structures crystallize in thermally stable rhombohedral symmetry. But, members of $A_3M_2(PO_4)_3$ family (where A=Li, Na and M=Cr, Fe) crystallize in monoclinic modification of $Fe_2(SO_4)_3$-type structure and show reversible structural phase transitions at high temperatures (d'Yvoire et al.,1983).

NASICON based phosphates are widely studied in past decades. But $LiTi_2(PO_4)_3$ is an interesting system because of its high ion conductivity at room temperature. The $Na_3Cr_2(PO_4)_3$ and $Li_3Fe_2(PO_4)_3$ are intriguing due to its structural peculiarity. These materials crystallize in structurally unstable phase by conventional synthesis technique. Since, $Na_3Cr_2(PO_4)_3$ and $Li_3Fe_2(PO_4)$ systems are not stable at the room temperature phase, a chemical synthesis technique of solution combustion is explored. In the present work we have achieved a stable phase through solution combustion technique and electrical properties are investigated and results are reported. The $LiTi_2(PO_4)_3$ and $Li_3Fe_2(PO_4)_3$ systems used as electrolytes in solid state batteries and $Na_3Cr_2(PO_4)_3$ system used in is sodium sensors. High energy ball milling technique can control the crystallite size through milling duration. In $LiTi_2(PO_4)_3$ system, milling is performed for various duration to study the effect of crystallite size on electrical conductivity.

To overcome the shortcomings in the conventional synthesis of NASICON, high-energy ball milling and solution combustion technique are explored. Correlation between mobile ion conduction and phase symmetry in NASICONs is explored in this study. Present chapter deals with the structure and electrical properties of important family of NASICONs like:

i. $LiTi_2(PO_4)_3$ and $Li_{1.3}Al_{0.3}Ti_{1.7}(PO_4)_{2.9}(VO_4)_{0.1}$ synthesized by high energy ball-milling.
ii. $A_3M_2(PO_4)_3$ (A=Li, Na and M=Cr, Fe) synthesised by solution combustion technique.

Characterization techniques like X-ray powder diffraction (XRD), Fourier-transform infrared spectroscopy (FT-IR), thermogravimetry and differential thermal analysis (TG-DTA) *etc* are exploited for structural confirmation of the synthesized material. Microscopy of the surface is analyzed using scanning electron microscope (SEM) and transmission electron microscope (TEM). UV-vis spectroscopy is used for confirmation of the electronic state of the transition elements and Kramers-Kronig test is performed for confirming the quality of measured electrical parameters. Transport number is measured by Wagner polarization technique. The electrical relaxation parameters are investigated in the frequency range 10Hz-25MHz at different temperatures using broadband dielectric spectrometer. Magnetic behavior of the material is investigated by vibrating sample magnetometer (VSM). In general, complex impedance, admittance, permittivity and modulus forms are used for representation of different electrical parameters. Present chapter uses impedance/dielectric spectroscopy technique for the electrical characterization of mobile ions.

2. Experimental details

Microcrystalline material is prepared by the conventional solid-state reaction of the stoichiometric mixture of Li_2CO_3 (Himedia, 99.0%), $NH_4H_2PO_4$ (Himedia, 99.0%), TiO_2 (LR grade, 98.0%), Al_2O_3 (Himedia, 99.0%) and V_2O_5 (Himedia, 99.0%). Overall reaction for the formation of $LiTi_2(PO_4)_3$ and $Li_{1.3}Ti_{1.7}Al_{0.3}(PO_4)_{2.9}(VO_4)_{0.1}$ [LATPV$_{0.1}$] are given as:

$$0.5Li_2CO_3+2TiO_2+3NH_4H_2PO_4 \xrightarrow{\Delta} LiTi_2(PO_4)_3+3NH_3+0.5CO_2+4.5H_2O$$

$$0.65Li_2CO_3+1.7TiO_2+0.15Al_2O_3+2.9NH_4H_2PO_4+0.05V_2O_5 \xrightarrow{\Delta} Li_{1.3}Ti_{1.7}Al_{0.3}(PO_4)_{2.9}(VO_4)_{0.1} + 0.65CO_2 + 2.9NH_3 + 4.35H_2O$$

Various steps involved in the synthesis of microcrystalline materials are:

i. Stoichiometric amounts of starting reagents were ground in an agate mortar for 45minutes.
ii. The mixture is placed in a silica crucible and slowly heated in an electric furnace up to 523K. Further, the temperature is increased to 623K and held at this temperature for 6h in order to ensure the total decomposition of the initial reagents.
iii. After cooling the mixture to room temperature, it is again ground for 45min in an agate mortar and pellets of 10mm diameter and 1-1.5mm thickness was formed. Further pellets were heat treated at 923K for 6h. Heating procedure remains the same for both the systems till this stage.
iv. Further, $LiTi_2(PO_4)_3$ pellets were calcined at 1073K for 36h followed by sintering at 1223K for 2h. In the meanwhile, the pellets of $Li_{1.3}Ti_{1.7}Al_{0.3}(PO_4)_{2.9}(VO_4)_{0.1}$ is calcined at 1073K for 48h followed by sintering at 1323K for 4h.

Crystallites of smaller size materials are prepared through conventional solid-state reaction of the ball-milled stoichiometric mixture. The mixture is heated at 623K before ball-milling to remove the gases and water content. This minimizes sticking property of the mixture to the vial and balls. The tungsten carbide vial and balls were used for high energy milling; the typical ball to powder mass ratio is kept at 5:1 throughout the milling. The rotation speed is kept at 300rpm, each cycle comprised of 2h run followed by 30minutes pause, and these cycles were repeated. Milling is carried out in an ethanol medium in case of $Li_{1.3}Ti_{1.7}Al_{0.3}(PO_4)_{2.9}(VO_4)_{0.1}$, which acts as a surfactant to decrease the agglomeration and helps to reduce the heat produced while milling. The powder obtained after milling was made into pellets and further heat treatments were applied from 923K to 1223K for $LiTi_2(PO_4)_3$, and 923K to 1323K for $Li_{1.3}Ti_{1.7}Al_{0.3}(PO_4)_{2.9}(VO_4)_{0.1}$ with the same duration as the microcrystalline sample. In this study, material is sintered at a temperature lower than the conventional ceramic route. Even though, the sintering temperature is low, long hours of sintering are performed to obtain the required density for samples. Low temperature sintering is applied to maintain the nanocrystalline nature of the samples.

Self propagating solution combustion synthesis is a rapid and energy saving technique that works on the principle of decomposition of an oxidizer, metal nitrate, in the presence of fuel/complexing agent . The $Na_3Cr_2(PO_4)_3$ using glycine in 1:1fuel ratio ($Na_3Cr_2(PO_4)_3$-G1:1) is prepared from $NaNO_3$ and $Cr(NO_3)_3.9H_2O$. Stoichiometric amount of the metal nitrates and glycine (NH_2-CH_2COOH) were mixed with distilled water in 1:1 molar ratio. The $NH_4H_2PO_4$ dissolved in distilled water is added to this mixture to form homogenous solution. Slow evaporation of the homogenous solution produced thick viscous gel. Further heating resulted in flame, producing voluminous powder named as-prepared material. Over all reaction for the formation of $Na_3Cr_2(PO_4)_3$-G1:1 is calculated as:

$$3NaNO_3 + 2Cr(NO_3)_3.9H_2O + 3NH_4H_2PO_4 + 8NH_2\text{-}CH_2COOH + 5O_2 \xrightarrow{\Delta} Na_3Cr_2(PO_4)_3 + 10N_2 + 16CO_2 + 47H_2O$$

In the case of glycine-nitrate combustion, primarily N_2, CO_2, and H_2O were evolved as gaseous products. As-prepared material is in amorphous phase and further heating at 800°C produced the pure $Na_3Cr_2(PO_4)_3$ phase. To understand the effect of fuel molar ratio on physical and electrical properties; glycine, urea and citric acid were used in 1:1, 1:2 and 1:3 molar ratios for the synthesis of $Na_3Cr_2(PO_4)_3$.

The Fe^{3+} based NASICON materials were synthesized using citric acid: ethylene glycol mixture (CA:EG). The metal cations were complexed by citric acid ($C_6H_8O_7$) and pH of the resultant solution is adjusted in the range 7-8 using ammonia solution. This solution is kept under constant stirring and $NH_4H_2PO_4$ is added to it. After proper stirring, ethylene glycol is added to this solution by maintaining the molar ratio with citric acid at 1:1. The homogenous solution is heated further and the as-prepared material is formed. Further calcination at 800°C resulted in pure phase. Objective of the present investigation is to synthesize nanocrystalline materials by a unique combination of citric acid (as complexing agent) and ethylene glycol (as polymerizing agent). In the presence of ethylene glycol, esterification (reaction between alcohol and acid) resulted in the formation of gel. The $Li_3Fe_2(PO_4)_3$ is also prepared using glycine in 1:2 molar ratio.

3. Results and discussion

3.1 X-ray powder diffraction analysis

X-ray patterns are not recorded in very low quality; it is collected using Philips X'pert pro-diffractometer with Bragg-Brentano geometry in θ–θ configuration. The monochromatic Cu-K_α radiation of wavelength, λ = 1.5406Å is used. The pattern is recorded in the 2θ range 5°-75° with step size of 0.02° and the step scan of 0.50 seconds. Figs. 1(a)-(b) show XRD patterns of the microcrystalline and 40h ball-milled $LiTi_2(PO_4)_3$ sintered at 1073K. Peaks in the diffraction pattern correspond to the rhombohedral phase but, minor phase of TiP_2O_7 are observed due to Li loss in high temperature sintered material [Aono et al.,19984 & Wong et al., 1998]. Fig.1(c) shows XRD pattern of microcrystalline, 22h and 55h ball-milled $Li_{1.3}Ti_{1.7}Al_{0.3}(PO_4)_{2.9}(VO_4)_{0.1}$ material. Lattice parameters are calculated using UNITCELL software (Unit-Cell software,1995), ball-milling decreases lattice parameters and unit cell volume of $LiTi_2(PO_4)_3$ (Delshad et al., 2009 & Hamzaoui et al., 2003). But, lattice parameters increase for $Li_{1.3}Ti_{1.7}Al_{0.3}(PO_4)_{2.9}(VO_4)_{0.1}$ with ball-milling (Prithu et al., 2009) as given in Table 1. The line broadening in XRD pattern occurs due to the simultaneous change in crystallite size and strain effects (Savosta et al., 2004), because high energy ball-milling introduces considerable strain in the material. The strain resulted in broadening the XRD peak and shifting the peak positions towards the higher 2θ values.

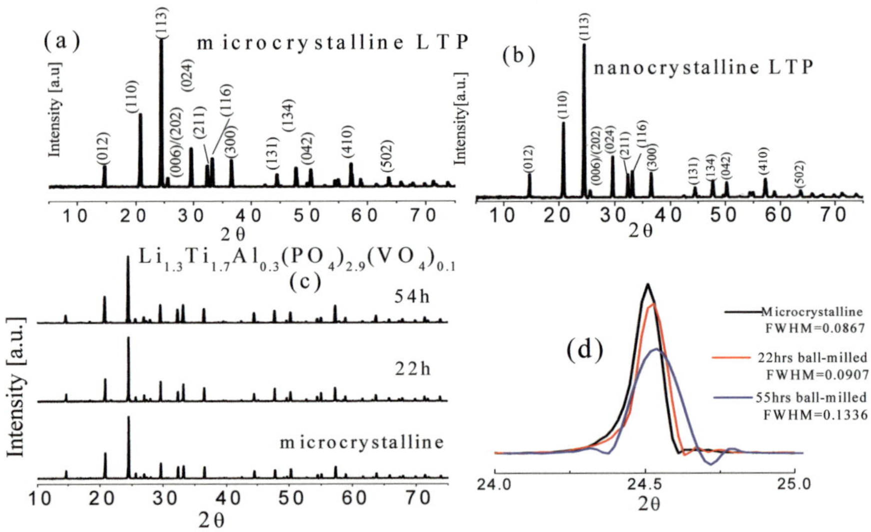

Fig. 1. X-ray powder diffraction patterns of (a) microcrystalline $LiTi_2(PO_4)_3$ (b) nanocrystalline $LiTi_2(PO_4)_3$ (c) $Li_{1.3}Ti_{1.7}Al_{0.3}(PO_4)_{2.9}(VO_4)_{0.1}$ microcrystalline, 22h and 55h ball-milled material and (d) Full width at half maximum of maximum intensity peak of microcrystalline, 22h and 55h ball-milled $Li_{1.3}Ti_{1.7}Al_{0.3}(PO_4)_{2.9}(VO_4)_{0.1}$.

Williamson and Hall (Williamson & Hall, 1953) developed a model to separate the size and strain effects in broadening the XRD peaks and the model is given by:

$$B\cos\theta = K\lambda/D + 4\varepsilon\sin\theta \quad (1)$$

where, B is the full width at half maximum (FWHM) of XRD peaks, K is the Scherrer constant, D is the crystallite size, λ is the wavelength of X-ray, ε is the micro-strain in the lattice and θ is the Bragg angle. For Gaussian X-ray profiles, B can be calculated as:

$$B^2=B_m^2-B_s^2 \quad (2)$$

where, B_m is the FWHM of the material and B_s is the FWHM of a standard sample; silicon is chosen as the standard for calculation of instrumental parameters. Linear extrapolation of the plot of Bcosθ *vs* 4sinθ gives average crystallite size from the intercept, Kλ/D and the slope gives micro-strain. The strain contribution in Eq. (1) is negligible for the crystallite size calculation of microcrystalline materials. Micro-strain and average crystallite size of $LiTi_2(PO_4)_3$ and $Li_{1.3}Ti_{1.7}Al_{0.3}(PO_4)_{2.9}(VO_4)_{0.1}$ are listed in Table 1.

Ball-milling induces strain in the crystal lattice and decreases the average crystallite size to 70nm for 40h ball-milled $LiTi_2(PO_4)_3$ material. Milling reduces the average size of crystallites to nanometer range and long hours of ball-milling lead to the formation of an amorphous state (Yamamoto et al., 2004 & Nobuya et al., 2005). Hence, sintering at high temperature after ball-milling resulted in the formation of nanocrystallites instead of microcrystalline material. XRD pattern gradually broadens and the particle size decreases with milling duration, which is clear from the FWHM of highest intensity peaks of ball-milled $Li_{1.3}Ti_{1.7}Al_{0.3}(PO_4)_{2.9}(VO_4)_{0.1}$ as given in Fig. 1(d). The nanocrystalline nature of the ball-milled materials is evident from the broadened XRD peak and there is decrease in peak intensity as compared to the microcrystalline material.

$LiTi_2(PO_4)_3$					
	Average crystallite size	Micro-strain	Unit cell parameters		
			a[A°]	c[A°]	V[A°]^3
Micro-crystalline	(0.23±0.01)μm	(0.05±0.001)%	8.514(9)	20.857(2)	1309.633(0)
Nano-crystalline	(70.14±0.07)nm	(0.36±0.05)%	8.495(9)	20.719(5)	1295.156(6)
$Li_{1.3}Ti_{1.7}Al_{0.3}(PO_4)_{2.9}(VO_4)_{0.1}$					
Micro-crystalline	(1.60 ±0.49)μm	(0.02±0.003)%	8.500(9)	20.819(6)	1302.958(1)
22h ball-milled	(86.62±0.27)nm	(0.29±0.04)%	8.504(1)	20.825(2)	1304.303(6)
55h ball-milled	(60.86±0.34)nm	(0.62±0.06)%	8.512(9)	20.845(0)	1308.254(0)

Table 1. Average crystallite size, micro-strain and unit cell parameters of microcrystalline and nanocrystalline $LiTi_2(PO_4)_3$ and $Li_{1.3}Ti_{1.7}Al_{0.3}(PO_4)_{2.9}(VO_4)_{0.1}$ materials.

The $Na_3Cr_2(PO_4)_3$ is synthesised using glycine, urea and citric acid in 1:1,1:2 and 1:3 molar ratios by solution combustion technique. The $Na_3Cr_2(PO_4)_3$ synthesized through conventional ceramic route is reported to exhibit two main structural phase transitions at 138°C and 166°C, before the stable rhombohedral symmetry is attained at high temperature (d'Yvoire et al.,1983). Fig. 2(a) shows the powder XRD patterns of $Na_3Cr_2(PO_4)_3$-G1:1, $Na_3Cr_2(PO_4)_3$-G1:2 and $Na_3Cr_2(PO_4)_3$-G1:3 pellets sintered at 900°C. The $Na_3Cr_2(PO_4)_3$, that are synthesised using citric acid in all molar ratios and urea in 1:3 molar ratio, are crystallized in mixed phase. Hence, further studies related to these compositions are not discussed in this chapter.

d'Yvoire *et al.*, reported the monoclinic symmetry (α-form) of $Na_3Cr_2(PO_4)_3$, at the room temperature, where, Na^+ ions are ordered at M1 site (d'Yvoire et al.,1983). The reversible phase transitions in conventionally synthesized $Na_3Cr_2(PO_4)_3$ are: α ↔ α' at 75°C; α'→β at 138°C and β↔γ at 166°C. In the high temperature γ-form of rhombohedral symmetry, Na^+ ions are distributed in M1 and M2 sites in the disordered manner. The temperature dependent XRD studies showed that α→γ phase transitions are associated with slight changes in the crystal lattice. Peaks in the DTA curve are not completely separable for α'↔β and β↔γ transitions, but it forms relatively broad endo or exothermic effect from 120°C to 178°C with two maxima. Change of slope in the Arrhenius plot around 75°C and increase in conductivity about 140°C are attributed to α↔α' and α'↔β transitions respectively. The β↔γ transition is associated with the decrease in activation energy (d'Yvoire et al., 1983).

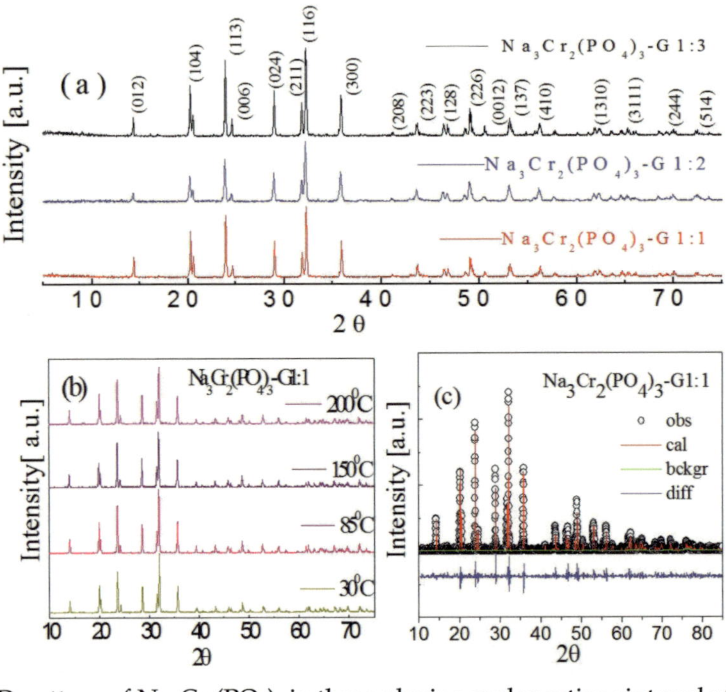

Fig. 2. (a) XRD pattern of $Na_3Cr_2(PO_4)_3$ in three glycine molar ratios sintered at 900°C **(b)** XRD patterns of $Na_3Cr_2(PO_4)_3$-G1:1 at 30°C, 85°C,150°C and 200°C **(c)** Rietveld refinement of $Na_3Cr_2(PO_4)_3$-G1:1 with observed, calculated and difference patterns.

While, nanocrystalline $Na_3Cr_2(PO_4)_3$ synthesized in the present study, is crystallized in thermally stable rhombohedral symmetry (JCPDS reference code: 01-084-1203). The XRD patterns are indexed and all reflections are from the rhombohedral phase. This type of structural modification is common in materials synthesized by the various chemical routes. In order to confirm the structural stability of $Na_3Cr_2(PO_4)_3$, XRD patterns are recorded at 30°C, 85°C, 150°C and 200°C. High temperature XRD patterns match well with the room temperature pattern and do not show any structural change with the temperature as shown in Fig. 2(b) for $Na_3Cr_2(PO_4)_3$-G1:1. The Rietveld refinement of room temperature XRD pattern of $Na_3Cr_2(PO_4)_3$-G1:1, is performed using GSAS computer package (Toby, 2001 & Larson, 1994) to confirm the crystal system. The Fig. 2(c) shows the Rietveld refinement, where symbol shows the experimental data collected in the slow scan mode, calculated and difference patterns are in solid lines with different colours. Refinement is performed based on rhombohedral crystal system in $R\bar{3}c$ space group. Initially, the parameters like zero shift, FWHM, background, scale factor and pseudo-Voigt coefficient are refined. Then lattice parameters, atomic positions of Cr, P and O are refined in 12c(0,0,z), 18e(x,0,1/4), and 36f(x,y,z) wyckoff positions respectively. Na^+ ions are assumed to occupy M1 and M2 sites partially; whose wyckoff positions are 6b(0,0,0) and 18e(x,0,1/4) respectively. The results of Rietveld refinement are given in Table 2. From these results, it is confirmed that in solution combustion synthesised $Na_3Cr_2(PO_4)_3$, Na^+ ions are distributed in M1 and M2 sites at the room temperature itself. Hence, this material does not show structural changes with temperature.

Atom	Site	Wyckoff position			B_{iso}	Occupancy
		x	y	z	$[A°]^2$	
Na(1)	6b	0.0000	0.0000	0.0000	1.448	0.84(1)
Na(2)	18e	0.655(3)	0.0000	0.2500	1.409	0.65(2)
Cr	12c	0.0000	0.0000	0.147(2)	1.551	1.000
P	18e	0.291(9)	0.0000	0.2500	2.224	1.000
O (1)	36f	0.181(5)	-0.039(7)	0.193(5)	3.479	1.000
O (2)	36f	0.199(3)	0.166(1)	0.0894	1.453	1.000

Table 2. Results of Rietveld refinement of $Na_3Cr_2(PO_4)_3$-G1:1. Atomic and isotropic displacement factors obtained from the refinement are provided below.

R_p =30.51(%), R_{wp} = 42.33(%), χ^2 = 3.258

Another member of the NASICON family, that shows structural phase transition is $Li_3Fe_2(PO_4)_3$. The $Li_3Fe_2(PO_4)_3$ synthesised by ceramic route is crystallized in $Fe_2(SO_4)_3$-type monoclinic symmetry and exhibited reversible structural phase transitions below 350°C, that are not completely separated (d'Yvoire et al.,1983). Its XRD patterns do not show any modifications due to structural phase transitions, implying the Li^+ ion distribution or ordering, rather than ordering of the networks. d'Yvoire *et al.*, and Bykov *et al.*, (Bykov,1990) showed that the monoclinic $Li_3Fe_2(PO_4)_3$ transforms reversibly to the orthorhombic phase upon heating above 270°C, due to progressive breaking of long-range ordering of Li^+ ions in the interstitial space. The $Fe_2(SO_4)_3$-type phase generally crystallize in two symmetries: (i) orthorhombic (Pcan) of highest symmetry and (ii) primitive monoclinic ($P2_1/n$) symmetry (Mineo, 2002).

In the present study, $Li_3Fe_2(PO_4)_3$ is synthesized by solution combustion technique using different fuels *i.e.*, glycine in 1:2 molar ratio and citric acid: ethylene glycol mixture in 1:1 molar ratio. Both of these $Li_3Fe_2(PO_4)_3$ is crystallized as mixture of monoclinic ($P2_1/n$) and orthorhombic (Pcan) symmetry. Due to sintering in air, XRD patterns showed presence of minor phases of $LiFeP_2O_7$ that crystallized in monoclinic symmetry. Fig. 3(a) shows XRD patterns of $Li_3Fe_2(PO_4)_3$, sintered at 900°C, synthesized using glycine. In the Figs. 3(a)-(c), black and red colour indexes are reflections from monoclinic and orthorhombic symmetry respectively. The violet colour index shows reflections due to $LiFeP_2O_7$ phase. In contradiction with the conventional synthesis process, solution combustion technique crystallized the material as a mixture of room temperature and high temperature phases. In the high temperature orthorhombic phase, alkali ions distribute disorderly in the available sites; hence the structural phase transitions are absent in the investigated $Li_3Fe_2(PO_4)_3$ material.

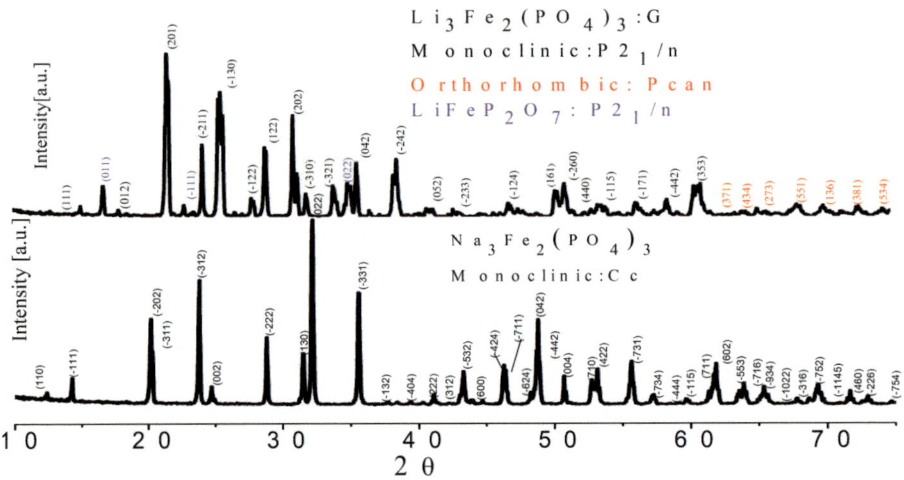

Fig. 3. XRD patterns of **(a)** $Li_3Fe_2(PO_4)_3$-G and (b) $Na_3Fe_2(PO_4)_3$ sintered at 900°.

The conventionally synthesised $Na_3Fe_2(PO_4)_3$ (d'Yvoire et al.,1983) showed two reversible phase transitions: (i) transition from monoclinic (C2/c) symmetry, $\alpha \leftrightarrow \beta$ below 368K and (ii) monoclinic to rhombohedral, $\beta \leftrightarrow \gamma$ at 418K, where, γ-phase ($R\bar{3}c$) is the stable symmetry. The monoclinic symmetry contains two formula units, *i.e.*, Z=2 and in this frame-work, Na^+ ions occupy three different sites.

In the present study, $Na_3Fe_2(PO_4)_3$ is synthesized by solution combustion technique using citric acid: ethylene glycol mixture in 1:1 molar ratio. The material is crystallized in monoclinic symmetry of Cc space group without an impurity phase. Fig. 3 shows the room temperature XRD pattern of $Na_3Fe_2(PO_4)_3$ sintered at 910°C. The high temperature XRD and

DTA studies confirmed the structural stability of solution combustion synthesised $Na_3Fe_2(PO_4)_3$. XRD patterns of $Na_3Fe_2(PO_4)_3$ is recorded at 30°C, 110°C, 300°C and 500°C. The high temperature XRD patterns match well with the room temperature pattern and do not show any structural change with temperature. Table 3 provides the crystal system and physical parameters of NASICON materials investigated in the present study.

Material	Crystal system	Space group	Lattice parameters				Crystallite size [nm]
			a [A°]	b [A°]	c [A°]	Volume [A°3]	
$Na_3Cr_2(PO_4)_3$-G1:1	Rhombohedral	$R\bar{3}c$	8.637(4)	8.637(4)	21.615(1)	1396.54(1)	31.29
$Na_3Cr_2(PO_4)_3$-G1:2			8.656(7)	8.657(4)	21.675(6)	1406.71(6)	34.02
$Na_3Cr_2(PO_4)_3$-G1:3			8.642(1)	8.642(1)	21.601(5)	1397.18(2)	39.39
$Na_3Cr_2(PO_4)_3$-U1:1			8.651(2)	8.651(2)	21.631(7)	1402.08(3)	44.50
$Na_3Cr_2(PO_4)_3$-U1:2			8.663(5)	8.663(5)	21.648(7)	1407.17(8)	55.45
$Li_3Fe_2(PO_4)_3$-G	Monoclinic and Orthorhombic	$P2_1/n$ and Pcan	8.597(3)	12.148(5)	8.635(4)	901.91(8)	55.43
$Li_3Fe_2(PO_4)_3$-CA: EG			8.578(1)	11.975(8)	8.734(1)	897.25(1)	54.60
$Na_3Fe_2(PO_4)_3$	Monoclinic	Cc	8.735(8)	12.138(5)	8.798(4)	932.97(8)	40.12

Table 3. The crystal system and physical parameters of NASICON materials

3.2 FT-IR analysis

FT-IR is one of the most general spectroscopic techniques used to identify the functional groups in materials. It is an important and popular tool for structural exposition and compound identification. The FT-IR spectra of NASICON materials are dominated by intense, overlapping intramolecular PO_4^{3-} stretching modes (v_1 and v_3) that range from 1300 to 700cm^{-1} (Corbridge and Lowe, 1954) . In most of the cases, experimentally measured vibrations are divided into internal and external modes. The internal vibrations consist predominantly of intramolecular stretching and bending motions of the PO_4^{3-} anions and are usually described in terms of the fundamental vibrations of the free anion (*i.e.,* v_1–v_4). Bands between 650 and 400cm^{-1} are attributed to the harmonics of deformation of O-P-O angle (v_2 and v_4 modes) (Rao, 2001). Bands in the region 580cm^{-1} are attributed to the asymmetric bending vibrational modes of O-P-O units (Sayer & Mansingh, 1972). The region 931–870cm^{-1} is assigned to PO_4^{3-} ionic group vibration (Rulmont, 1991). The entire region down to 400cm^{-1} is dominated by vibrations of PO_4 tetrahedra group. Stretching vibrations of P–O–P bond are identified in the region 700-758cm^{-1} (Alamo & Roy, 1998; Kravchenko et al. 1992; Rougier, 1997). Further, FT-IR spectra show weak peak of carbonates in the region 1400-1600cm^{-1}.

The FT-IR absorption bands of ball-milled $Li_{1.3}Ti_{1.7}Al_{0.3}(PO_4)_{2.9}(VO_4)_{0.1}$ in the range 1600cm^{-1}-400cm^{-1} are shown in Fig. 4. The asymmetric stretching vibration of VO_4 tetrahedra is observed at 810-850cm^{-1} as broad band (Benmokhtar, 2007). In addition, oxygen atom in the VO_4 tetrahedra can form bond with Al atom which can lead to some asymmetry. The stretching modes of VO_4 in the IR spectra confirm the substitution of vanadium for phosphorus in PO_4 tetrahedra.

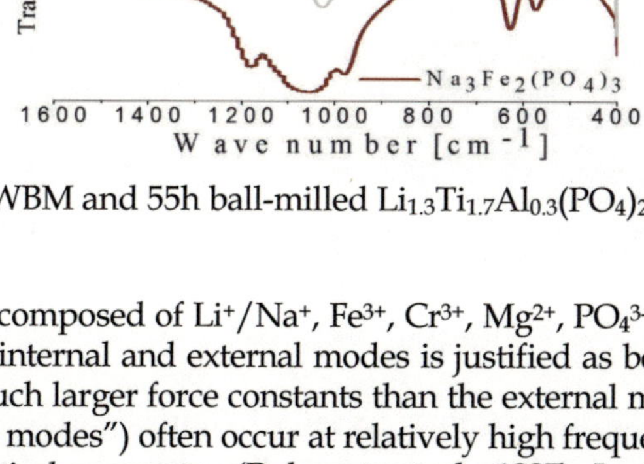

Fig. 4. FT-IR spectra of WBM and 55h ball-milled $Li_{1.3}Ti_{1.7}Al_{0.3}(PO_4)_{2.9}(VO_4)_{0.1}$, $Li_3Fe_2(PO_4)_3$ and $Na_3Fe_2(PO_4)_3$.

The external modes are composed of Li^+/Na^+, Fe^{3+}, Cr^{3+}, Mg^{2+}, PO_4^{3-} translations and pseudo-rotations. Separation of internal and external modes is justified as because the intramolecular PO_4^{3-} vibrations have much larger force constants than the external modes. The Li^+ translatory vibrations (Li^+ ion "cage modes") often occur at relatively high frequencies and mix with PO_4^{3-} bending modes of identical symmetry (Rulmont, et al. ,1997). In these vibrations, Li^+ ions undergo translatory motions in a potential energy environment, that is determined by the nearest neighbour oxygen atoms. Bands in the region 1227-185cm^{-1} of $Na_3Cr_2(PO_4)_3$ correspond to the interaction of P-O bond and adjacent Cr-O bond (Alamo & Roy, 1986).

3.3 SEM-EDS and TEM analysis

The crystallites in nanocrystalline $LiTi_2(PO_4)_3$ are agglomerated and its size distribution is not uniform due to dry milling (Puclin,1995). The quantitative chemical analysis is performed through EDS, but it cannot detect elements with atomic number less than four and hence Li metal cannot be detected by this technique. Surface morphology of $Li_3Ti_{1.7}Al_{0.3}(PO_4)_{2.9}(VO_4)_{0.1}$ material also shows agglomeration of crystallites in the ball-milled samples and the particle size decreases with milling duration as shown in Figs. 5(a)-(c). X-ray mapping is an imaging technique performed using X-ray. This analytical technique provides a high magnification image related to the distribution and relative abundance of elements within a given specimen. This technique is useful for: (i) identifying the location of individual elements and (ii) mapping the spatial distribution of specific elements and phases in the material surface. Figs. 5(d(ii)-(vi)) show X-ray dot mapping of the SEM image of the 55h ball-milled $Li_{1.3}Ti_{1.7}Al_{0.3}(PO_4)_{2.9}(VO_4)_{0.1}$ material shown in d(i). Elemental analysis shows peaks corresponding to Ti, Al, P and O elements present in the material. The inset table in Fig. 6(e) give weight and atomic percentage of elements present in 55h ball-milled $Li_{1.3}Ti_{1.7}Al_{0.3}(PO_4)_{2.9}(VO_4)_{0.1}$ material. The percentages of elements that are detected by instrument and calculated from molecular formula fall within the error.

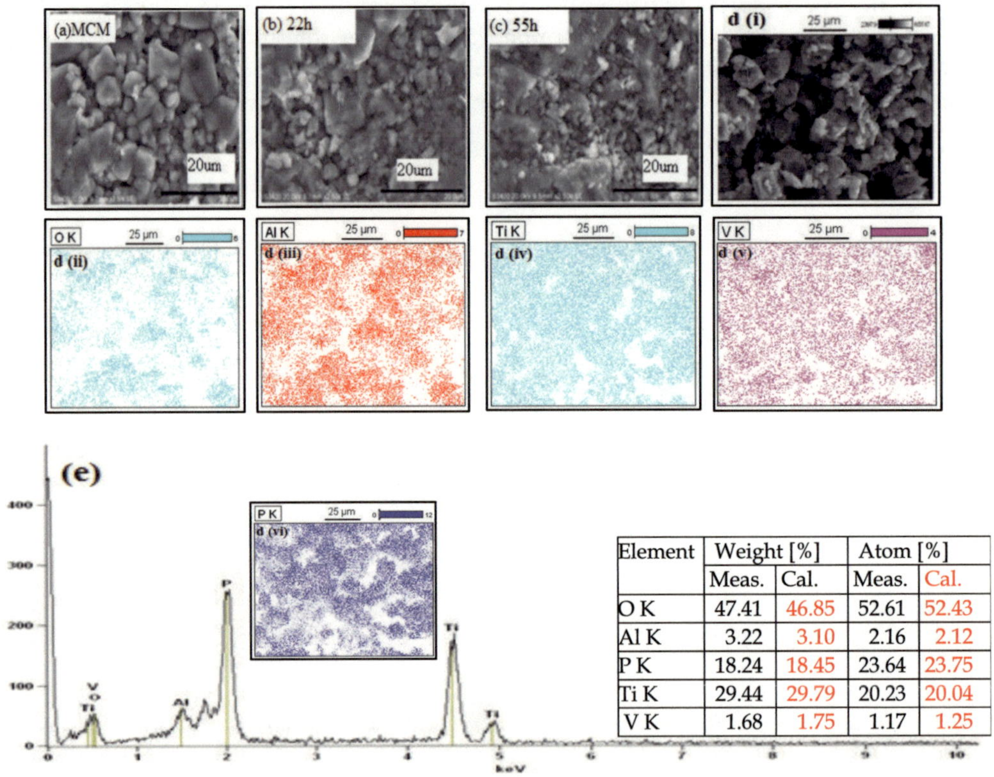

Element	Weight [%]		Atom [%]	
	Meas.	Cal.	Meas.	Cal.
O K	47.41	46.85	52.61	52.43
Al K	3.22	3.10	2.16	2.12
P K	18.24	18.45	23.64	23.75
Ti K	29.44	29.79	20.23	20.04
V K	1.68	1.75	1.17	1.25

Fig. 5. SEMs image of (a) microcrystalline (b) 22h ball-milled and (c) 55h ball-milled $Li_{1.3}Ti_{1.7}Al_{0.3}(PO_4)_{2.9}(VO_4)_{0.1}$ d(i) SEM image of 55h ball-milled material and (d(ii)-(vi)) show X-ray mapping of d(i) image and (e) EDS spectrum shows peaks corresponding to the elements present in 55hball-milled material and the inset table give atomic and weight percentage of the elements.

Figs. 6(a)-(c) are SEM images of sintered pellets of $Na_3Cr_2(PO_4)_3$, synthesized using glycine in different fuel/complexing agent molar ratios. The molar ratios affect product morphology and sinterability. It is evident from SEM images that, the crystallite's density decreases and agglomeration increases with the molar ratios. The surface morphology reveals that the particles are of submicron size.

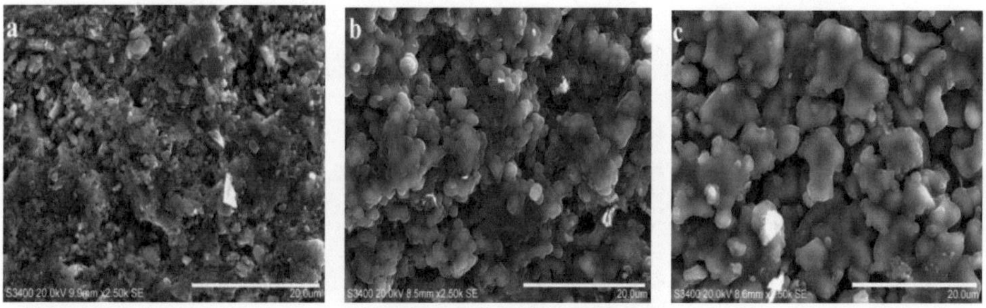

Fig. 6. The SEM images of sintered pellet of (a) $Na_3Cr_2(PO_4)_3$-G1:1 (b) $Na_3Cr_2(PO_4)_3$-G1:2 and (c) $Na_3Cr_2(PO_4)_3$-G1:3.

The nanocrystalline nature of the samples is confirmed from TEM images. The Figs. 7(a)-(b) show TEM images of sintered $Na_3Cr_2(PO_4)_3$-G1:1 and $Na_3Cr_2(PO_4)_3$-G1:3 materials. The agglomeration of nanometer sized crystallites is seen in TEM images. Fig. 7(c) is the diffraction pattern of the selected area from the microscopic image of $Na_3Cr_2(PO_4)_3$-G1:3 in Fig. 7(b). Due to strong association, the individually well separated microcrystals are not observable in the TEM images.

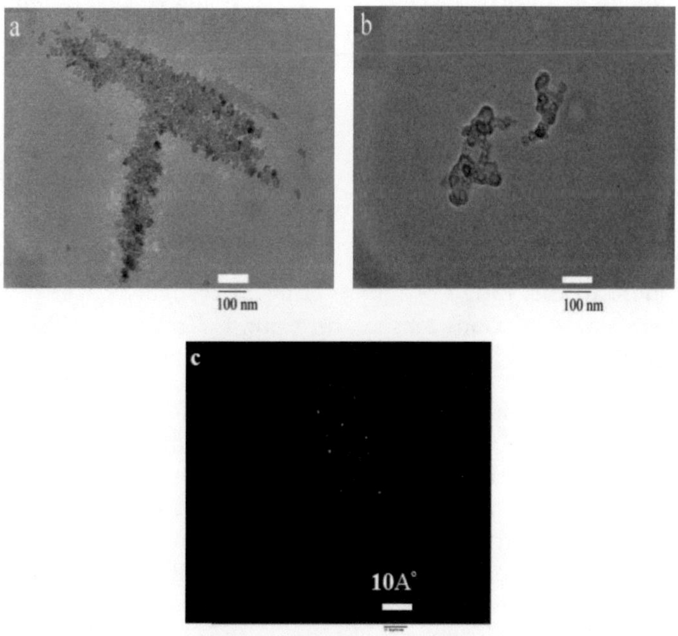

Fig. 7. TEM images of 900°C sintered (a) $Na_3Cr_2(PO_4)_3$-G1:1 (b) $Na_3Cr_2(PO_4)_3$-G1:3 and (c) Diffraction pattern of $Na_3Cr_2(PO_4)_3$-G1:3.

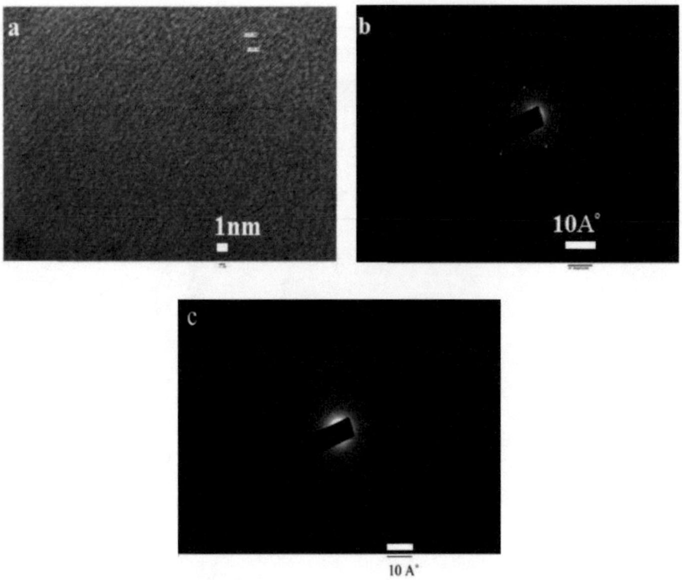

Fig. 8. (a) Lattice fringes of $Li_3Fe_2(PO_4)_3$-G in different orientations (b) and (c) show diffraction patterns from different (hkl) planes of $Li_3Fe_2(PO_4)_3$-G.

TEM image of $Li_3Fe_2(PO_4)_3$-G in Fig. 8(a) shows lattice fringes with different orientation. The lattice fringes with d-spacing of 1.98A° are identified in the images. In comparison with the XRD pattern, these fringes correspond to (-1, 2, 4) plane. Figs. 8(b)-(c) show the diffraction patterns from different planes of $Li_3Fe_2(PO_4)_3$-G. TEM image in Fig. 8(c) corresponds to 800°C sintered $Li_3Fe_2(PO_4)_3$ and the material is not well sintered at this temperature. The amorphous regions in the TEM image are due to less sintering.

3.4 Thermal analysis

Thermal studies include measurement of time dependence of material's temperature, while it is subjected to temperature-time variation. DSC measurements are also carried out for the phase transition analysis. But, DSC measurements are performed up to 500°C due to instrument limitation. In this range of temperature, investigated systems of $Na_3Cr_2(PO_4)_3$ and $Li_3Fe_2(PO_4)_3$ are stable at room temperature phase. To confirm the phase stability at higher temperature, DTA measurement is carried out.

Thermal and gravimetric analyses of as-prepared materials are carried out in the temperature range 40°C to 1000°C. Thermal study confirms the structural phase transition in the material and change in enthalpy of the products is calculated from the area of crystallization peak. TG-DTA curves of as-prepared $Na_3Cr_2(PO_4)_3$ in different fuel molar ratios are shown in Figs. 9(a)-(c). Out of the two exothermic peaks observed in DTA, the broad peak around 200-400°C corresponds to the decomposition of organic fuel/complexing agent and nitrates. The sharp peak between 740-780°C represents the crystallization process. The gravimetric plot shows significant weight loss in the temperature range 300°C to 740°C, that is due to the decomposition of organic intermediate and the crystallization process. Further weight loss between 740°C and 800°C is due to the formation of NASICON phase that is articulated in the DTA plot as sharp exothermic peak. The weight loss curve follows the same path for all materials, but the percentage of weight loss is more for higher molar ratios, due to the presence of more amount of carbonaceous residue. Absence of any additional peaks, in the DTA plot of the as-prepared material, ruled out the possibility of thermodynamical changes due to structural transition.

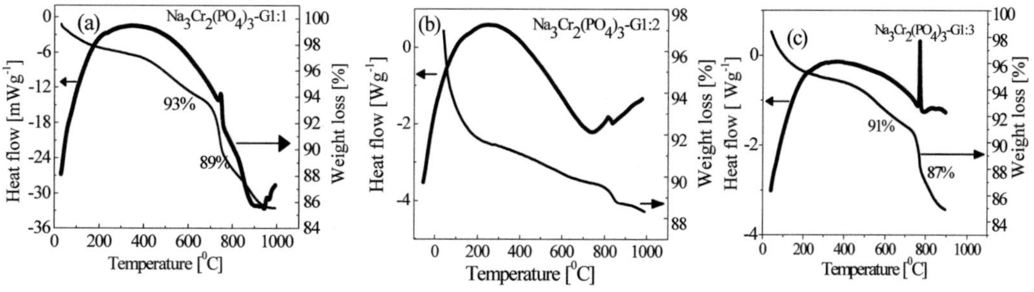

Fig. 9. Thermal and gravimetric plots of as-prepared (a) $Na_3Cr_2(PO_4)_3$-G1:1 (b) $Na_3Cr_2(PO_4)_3$-G1:2 and (c) $Na_3Cr_2(PO_4)_3$-G1:3.

The surface area and crystallite size are primarily decided by enthalpy or flame temperature of combustion process. The flame temperature depends on the nature of fuel/complexing agent and its molar ratio. Rapid evolution of large volume of gaseous products during combustion process dissipates heat, whereby limits the increase of temperature. This reduces the possibility of premature local partial sintering among the primary particles and helps in limiting the inter-particle contact. The crystallite size is decided mainly by two factors *i.e.,* adiabatic flame temperature and number of moles of gases released during combustion process. These two factors are more for higher fuel/complexing agent molar ratios. Higher values of flame temperature result in the formation of dense agglomerates that are disintegrated by the release of more amounts of gases (Hahn, 1990). The competition between flame temperature and number of moles of gases released decides the crystallite size. Crystallites of $Na_3Cr_2(PO_4)_3$-G1:1 are the smallest among the three fuel/ complexing agent ratios. Table 3 gives variation of crystallite size with molar ratios. In the present study, flame temperature has a major role than the number of moles of gases released, on controlling the crystallite size. DTA curves of $Li_3Fe_2(PO_4)_3$ prepared using different fuels/complexing agents are different due to the difference in the chemical decomposition of organic components. Table 4 provides crystallization temperature of investigated NASICONs obtained from the DTA plot. The crystallization temperature depends on nature of the fuel/complexing agent and its molar ratio.

DTA has been used to confirm the possible reversible structural phase transition in NASICON type materials. Fig. 10 shows the typical heating and cooling curve of $Na_3Cr_2(PO_4)_3$-G1:3 in the temperature range 40°C to 900°C (at the rate of 10°C/minute for both heating and cooling). The heating/cooling curves did not show exothermic/endothermic effect corresponding to phase transitions. This ruled out the possibility of structural phase transitions in $Na_3Cr_2(PO_4)_3$ and $Li_3Fe_2(PO_4)_3$ materials synthesized by solution combustion technique.

Material	Crystallization Temperature, [°C]
$Na_3Cr_2(PO_4)_3$-G1:1	743
$Na_3Cr_2(PO_4)_3$-G1:2	746
$Na_3Cr_2(PO_4)_3$-G1:3	771
$Na_3Cr_2(PO_4)_3$-U1:1	752
$Na_3Cr_2(PO_4)_3$-U1:2	805
$Li_3Fe_2(PO_4)_3$-G	864
$Li_3Fe_2(PO_4)_3$-CA:EG	885
$Na_3Fe_2(PO_4)_3$	819

Table 4. Crystallization temperatures obtained from the DTA plot.

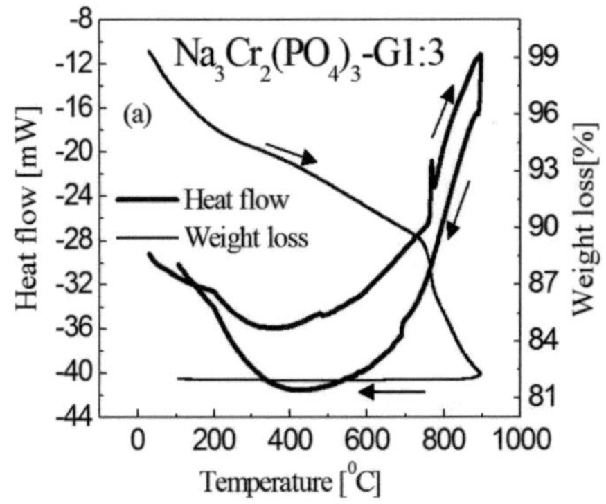

Fig. 10. Thermal analysis of as-prepared $Na_3Cr_2(PO_4)_3$-G1:3 (both heating and cooling curve).

3.5 Ultraviolet and visible absorption spectroscopy analysis

UV-vis spectroscopy is a tool for identifying valency (electronic) state of transition metals. The transition metals like Fe and Cr show variable valencies and can co-ordinate tetrahedrally and octahedrally. Each co-ordination state produces its own set of characteristic absorption bands in the visible and near UV range. These characteristic absorption bands are used to find skeleton co-valency of the material, which is related to the electronic contribution.

The UV-vis spectra of $Na_3Cr_2(PO_4)_3$ in three different glycine molar ratios has two absorption peaks as shown in Fig. 11(a). The $3d^3$ configuration of Cr^{3+} has a ^4F fundamental state with ^4P as the first excited state. The spin allowed transitions appeared at 670, 468 and 300nm are: ν_1:$^4A_{2g}(F) \rightarrow {}^4T_{2g}(F)$, ν_2:$^4A_{2g}(F) \rightarrow {}^4T_{1g}(F)$, ν_3:$^4A_{2g}(F) \rightarrow {}^4T_{1g}(P)$. Out of these three bands, ν_3 band appears occasionally [Stalhandske, 2000]. In $Na_3Cr_2(PO_4)_3$ material, Cr^{3+} does not show variable valency state and its contribution to the electronic part is negligible. The Fe^{3+} ions reveal absorption bands in the visible and near UV range. Both Fe^{2+} and Fe^{3+} ions can exist in tetrahedral and octahedral sites, and majority of Fe^{3+} ions are believed to occupy the tetrahedral network. The double absorption band at 340 and 380nm may be attributed to $^4D_5$ for ferric ion in tetrahedral state and absorption at 440nm is due to $^4G_5$ for ferric ion mostly in tetrahedral form. In addition to that, the absorption band at 560-580nm may be due to the presence of ferric ion in octahedral site (ElBatal et al., 1988; Bates & Mackenzie, 1962; Kurkjian & Sigety, 1968; Steele & Douglas, 1965; Edwards & Paul, 1972). The absorption band at 410-420nm of transition, $^3F_2 \rightarrow ^3P_4$, is related to the ferric ion in octahedral symmetry. $Li_3Fe_2(PO_4)_3$-CA:EG, $Li_3Fe_2(PO_4)_3$-G and $Na_3Fe_2(PO_4)_3$-CA:EG show absorption bands in the region of 420nm, 550nm and 720nm as in Fig. 11(b). These absorption peaks correspond to Fe^{3+} ions in octahedral state, this ruled out the presence of Fe^{2+} ions in the material. The present study concluded that, the dominating contribution to the total conductivity is from ions and the electronic part is negligible. The broad band in the region 200–400nm is due to the phosphate group and its location is independent of the nature of the cation.

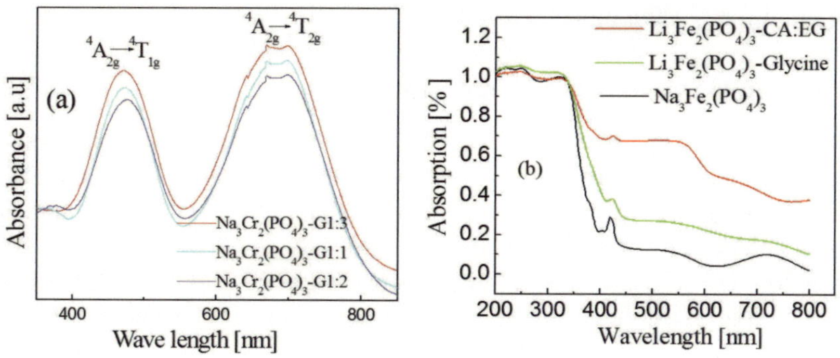

Fig. 11. UV-vis spectrum of (a) $Na_3Cr_2(PO_4)_3$-G1:1, G1:2 and G1:3 and (b) $Li_3Fe_2(PO_4)_3$-G, $Li_3Fe_2(PO_4)_3$-CA:EG and $Na_3Fe_2(PO_4)_3$.

3.6 Wagner polarization technique

The NASICON materials investigated in the present study contain transition elements like Cr and Fe. Due to the presence of variable valency states, these elements may contribute to the electronic part in the total conductivity. The electronic contribution is determined quantitatively by transport number measurement through Wagner polarization technique. The transport number is calculated from the instantaneous and steady state values of current obtained by dc polarization technique. The transport number of investigated materials in the present study is found to be approximately equal to one. Hence, the contribution of electronic part to total conductivity is negligibly small and this corroborates the results from VSM data.

3.7 K-K transformations

To validate the electrical microstructure of the material, like grain, grain-boundary and other external parameters such as electrode polarization, ac electrical parameters are plotted in the complex impedance formalism. Kramers-Kronig (K-K) relation is used to evaluate the quality of the measured impedance data. The K-K relations are true for complex impedance spectroscopic data that are linear, causal, and stable. Fig. 12(a) shows K-K fit to $Na_3Cr_2(PO_4)_3$-G1:3 at different temperatures and (b) shows K-K fit to $Na_3Cr_2(PO_4)_3$ in different glycine molar ratios at 393K. All these fits match well with the experimental data, implying good quality of the measured data. Kramers-Kronig fit to the complex impedance data is achieved through the software K-K test. Solid line shows K-K fit to the experimental data at different temperatures.

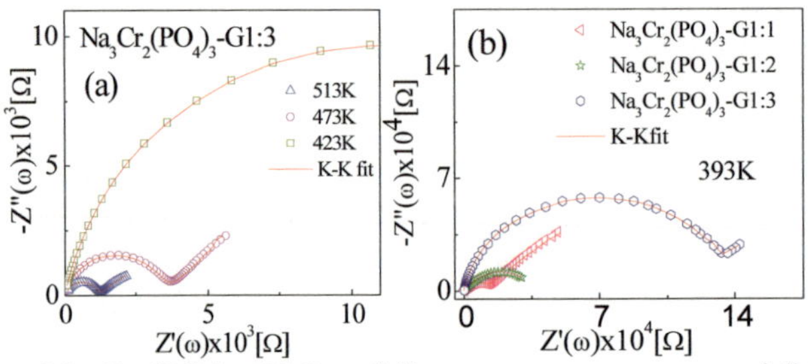

Fig. 12. (a) K-K fit to $Na_3Cr_2(PO_4)_3$-G1:3 at different temperatures and (b) K-K fit to $Na_3Cr_2(PO_4)_3$ in three glycine molar ratios at 393K.

3.8 Vibrating sample magnetometer analysis

In the present study, magnetization of NASICON materials are recorded over a range of field, using VSM at room temperature. Electronic contribution to the total conducivity is related to the co-existence of different electronic states. Generally, exchange interaction between equal valence ions is antiferromagnetic and interaction between ions with different valence states like Fe^{3+} ($3d^5$) and Fe^{2+} ($3d^6$) is ferromagnetic (Takano, 1981; Li, 1997). VSM measurement of the investigated NASICON materials as in Fig.13 show antiferromagnetic behaviour. This indicates that, the contribution to electronic conductivity is negligible in these materials.

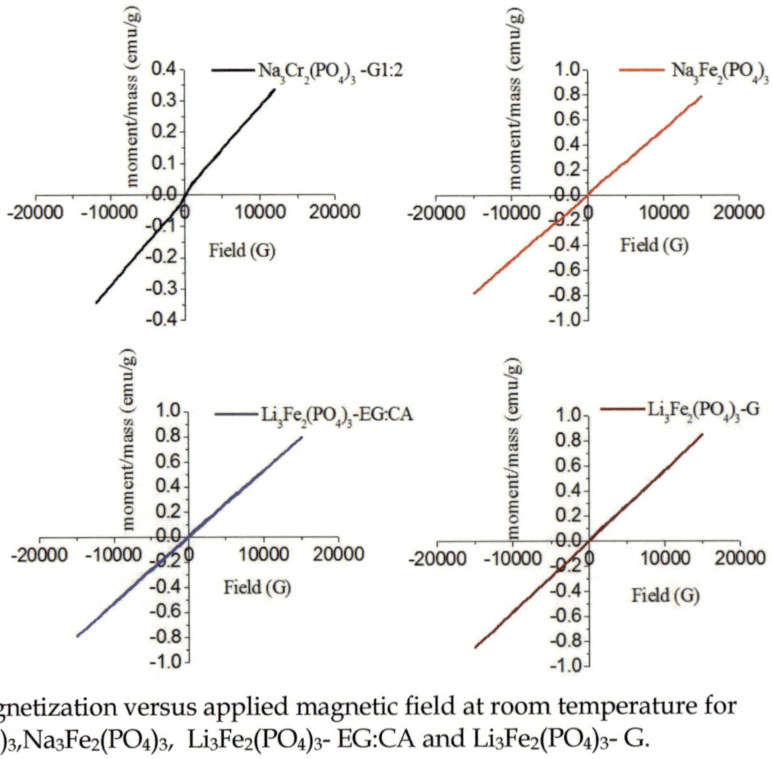

Fig. 13. Magnetization versus applied magnetic field at room temperature for $Na_3Cr_2(PO_4)_3$,$Na_3Fe_2(PO_4)_3$, $Li_3Fe_2(PO_4)_3$- EG:CA and $Li_3Fe_2(PO_4)_3$- G.

4. Impedance spectroscopy analysis

The real part, $Z'(\omega)$ and the imaginary part, $Z''(\omega)$ of the complex impedance, $Z^*(\omega)= Z'(\omega)-iZ''(\omega)$ are calculated from the measured G and C values as:

$$Z'(\omega)= G/(G^2+C^2\omega^2) \quad (3)$$

$$Z''(\omega)=C\omega/(G^2+C^2\omega^2) \quad (4)$$

where, $\omega=2\pi f$, f being the frequency in Hertz.

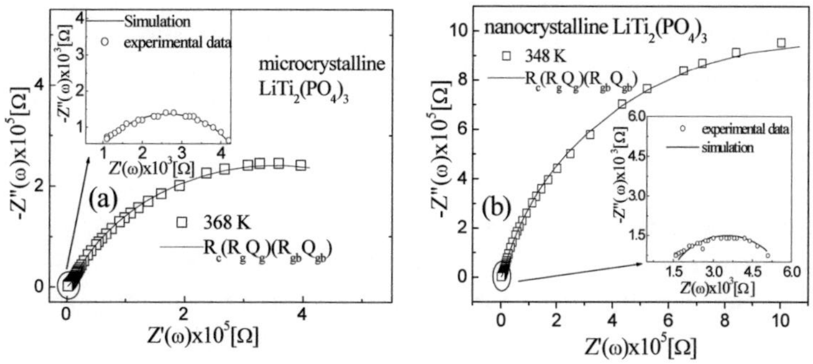

Fig. 14. The complex impedance spectra of (a) microcrystalline $LiTi_2(PO_4)_3$ at 368K and (b) nanocrystalline $LiTi_2(PO_4)_3$ at 348K. Inset of Fig. 14(a) and (b) shows grain part of the corresponding equivalent circuit and the continuous line is the simulation result.

The elements of an equivalent circuit model represent various (macroscopic) processes involved in the transport of mass and charge. Using NLLS techniques, all the parameters in the equivalent circuit are adjusted simultaneously, thus obtaining the optimum fit to the measured dispersion data. A more general NLLS-fit program based on the Marquardt algorithm has been used. The impedance parameters are obtained by fitting the data to an equivalent circuit using NLLS fitting procedure due to Boukamp [Boukamp, 1989; Mariappan & G. Govindaraj, 2004 & 2006).

Figs. 14 (a) and (b) show the complex impedance plane plot of microcrystalline $LiTi_2(PO_4)_3$ material at 368K and nanocrystalline $LiTi_2(PO_4)_3$ material at 348K. For both of these materials equivalent circuit model is the same throughout the temperature range from 309K to 388K. Equivalent circuit model consists of two depressed semi-circles, where the high frequency semi-circle is displaced from the origin. Since, the high frequency semi-circle is impeded by the low frequency one and effectively only one semi-circle can be visible in the complex impedance plane plot. The ratio of grain capacitance to the grain-boundary capacitance should be less than 10^{-3} for the appearance of two separate semi-circles in the complex impedance plane plot (Barsoukov & Macdonald, 2005; Mariappan & Govindaraj, 2005). Inset of Figs. 14(a) and (b) show the high frequency part in the complex impedance plane plot, where continuous line is the simulation result. Simulation clearly shows the grain semi-circle, which is not seen explicitly in the complex impedance representation of the equivalent circuit.

The impedance plane plots are depressed due to the distribution of relaxation times; a non-ideal capacitor or the CPE, Q, is used to explain the depressed semi-circle (Lakshmi et al., 2011,2011) Equivalent circuit of the impedance plane plots obtained using the Boukamp equivalent circuit analysis is found to be $R_c(R_gQ_g)(R_{gb}Q_{gb})$. Resistance of the electrolyte-electrode contact is R_c, which is characterized by the shift of the impedance arc from the origin. Constant phase elements, Q_g and Q_{gb} represent the grain and grain-boundary property of the sample. Grain resistance, R_g and the grain-boundary resistance, R_{gb} of the sample are obtained by right and left intercepts of the semi-circles with the real axis. R_g and R_{gb} are used to calculate the corresponding grain conductivity, σ_{dcg} and grain-boundary conductivity, σ_{dcgb}. The obtained equivalent circuit is the same for both the $LiTi_2(PO_4)_3$ samples, but with the different magnitudes of circuit parameters.

For both the samples, R_c variation is not consistent with temperature, R_g and R_{gb} of both the samples decrease with increase in temperature, Q_{gb} values increases with temperature, while, Q_g decreases. The grain conductivity at 309K ($\sigma_{dcg309K}$=1.82x10^{-6}Scm^{-1}) of the microcrystalline material is consistent with the reported room temperature value of 10^{-7}Scm^{-1} (Palani Balaya, 2006). At 388K, grain conductivity ($\sigma_{dcg388K}$=8.57x10^{-4}Scm^{-1}) of nanocrystalline material shows an order of magnitude jump (Lakshmi et al, 2009, 2011) compared to the microcrystalline material ($\sigma_{dcg388K}$=7.74x10^{-5}Scm^{-1}). This significant increase in the grain conduction resulted from the reduced crystallite size. High energy ball-milling introduces grain-boundaries in the material and its volume fraction is more in nanocrystalline material. The diffusion through grain-boundaries is much faster than the grain diffusion; hence large volume fractions of grain-boundaries play a dominant role in the ion conduction (Schoonman, 2003).

Figs. 15(a) and (b) show Arrhenius plot of grain and grain-boundary conductivity of the microcrystalline and nanocrystalline $LiTi_2(PO_4)_3$ material. The Arrhenius equation is given by:

$$\sigma_{dc}T=\sigma_0\exp(-E_\sigma/k_BT) \quad (5)$$

where, σ_{dc} is the dc conductivity, σ_0 is the pre-exponential factor, T is the temperature in Kelvin, E_σ is the activation energy for dc conduction and k_B is the Boltzmann's constant.

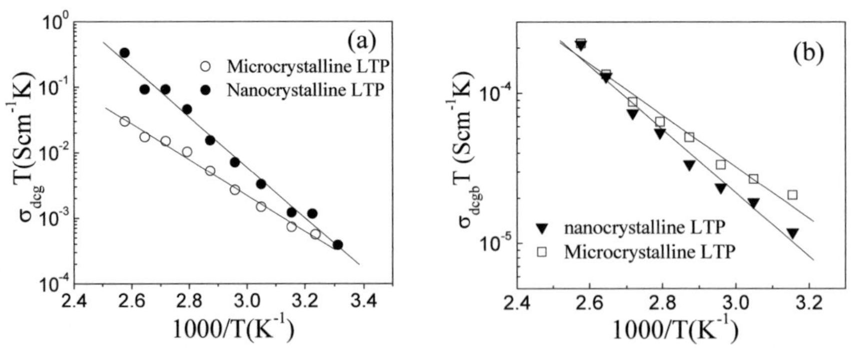

Fig. 15. Arrhenius plot of (a) grain and (b) grain-boundary conductivity of microcrystalline and nanocrystalline $LiTi_2(PO_4)_3$ material. Solid line represents best fit to the Eq. (5).

From the slope of the Arrhenius plots, grain and grain-boundary activation energies $E_{\sigma g}$ and $E_{\sigma gb}$ are calculated and are shown in Table 5. Increase in the grain conductivity of nanocrystalline material is due to the feasible conduction through grain-boundaries as its activation energy for grain-boundary conduction is less compared to the grain conduction (Mouahid, 2001). Even though the ball-milling decreases the crystallite size, its distribution is not uniform due to dry milling. The non-uniform size distribution and agglomeration are the causes of higher activation energy in nanocrystalline $LiTi_2(PO_4)_3$ material, in spite of its higher conductivity (Lakshmi et. al, 2009, 2011). These agglomerated crystallites are seen clearly in SEM images. Table 6 provides the charge carrier concentration, n_c, of microcrystalline and nanocrystalline $LiTi_2(PO_4)_3$ material, which authenticate that the ball-milling does not increase the carrier concentration.

$LiTi_2(PO_4)_3$	Activation energy for conduction through [eV]		Grain conductivity,[Scm^{-1}]	
	Grain, $E_{\sigma g}$	Grain-boundary, $E_{\sigma gb}$	σ_{dcg} at 298K	σ_{dcgb} at 388K
Microcrystalline	(0.54±0.02)	(0.34±0.02)	2.34x10-6	7.74x10-5
Nano-crystalline	(0.76±0.03)	(0.42±0.02)	1.28x10-6	8.57x10-4

Table 5. Activation energies and dc conductivity values of microcrystalline and nanocrystalline $LiTi_2(PO_4)_3$ materials.

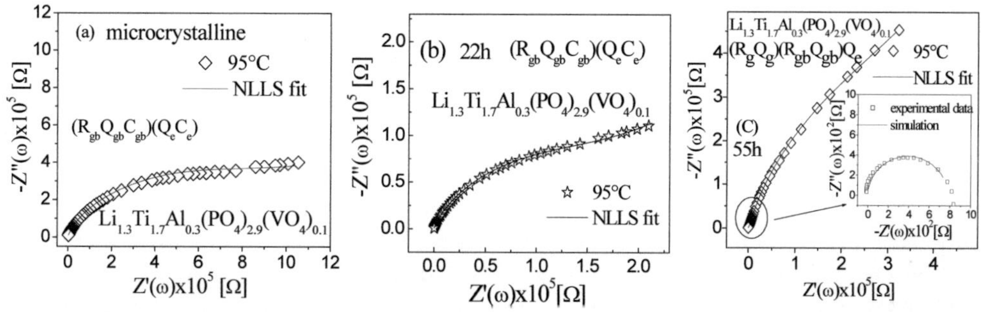

Fig. 16. Complex impedance spectra of (a) microcrystalline (b) 22h and (c) 55h ball-milled $Li_{1.3}Ti_{1.7}Al_{0.3}(PO_4)_{2.9}(VO_4)_{0.1}$ material at 95°C. In Fig. 16(c) inset shows simulation to the grain semi-circle.

The impedance plane plots of $Li_{1.3}Ti_{1.7}Al_{0.3}(PO_4)_{2.9}(VO_4)_{0.1}$ material shows obvious indication of blocking effect at the grain-boundaries and at the electrode-sample interface. Figs. 15(a)-(c) show impedance plots of microcrystalline and ball-milled materials. Equivalent circuit consists of series combination of a semi-circle associated to grain-boundary contribution and spike characterizing the electrode disparity at the low frequency part. The equivalent circuit representation is $(R_{gb}Q_{gb}C_{gb})$ up to 85°C and at higher temperatures it becomes $(R_{gb}Q_{gb})(Q_eC_e)$ for the microcrystalline material. In the case of 22h ball-milled material, the equivalent circuit representation is $(R_{gb}Q_{gb})(Q_eC_e)$ in the whole temperature range. Impedance plane plots of 55h ball-milled material show overlapped semicircles; in which the high frequency arc is attributed to the grain contribution.

Inset of Fig. 16(c) shows the high frequency part in the complex impedance plane plot where continuous line is the simulation result. Simulation clearly shows the grain semi-circle, which is not seen explicitly in the complex impedance plane representation. The high frequency studies are requisite to obtain the grain contribution of microcrystalline and 22h ball-milled material. Mechanical milling changes the capacitive contribution in such a way that in 55h ball-milled material, grain contribution is substantial within the frequency window (Lakshmi et al, 2009, 2011). Mechanical milling decreases the difference between the grain and grain-boundary capacitance values; which indicates relatively good connectivity between the grains.

Temperature [K]	Carrier concentration, n_c [cm^{-3}]	
	Microcrystalline	nanocrystalline
308	$2.01x10^{20}$	$1.09x10^{20}$
318	$2.36x10^{20}$	$7.77x10^{20}$
328	$2.16x10^{20}$	$1.05x10^{20}$
338	$2.12x10^{20}$	$7.41x10^{20}$
348	$2.13x10^{20}$	$7.56x10^{20}$
358	$1.98x10^{20}$	$5.27x10^{20}$
368	$1.95x10^{20}$	$3.59x10^{20}$
378	$2.17x10^{20}$	$4.65x10^{20}$
388	$1.84x10^{20}$	$5.12x10^{20}$
398	$1.76x10^{20}$	$3.75x10^{20}$

Table 6. Carrier concentartion of microcrystalline and nanocrystalline $LiTi_2(PO_4)_3$ materials over the temperature range 308K to 398K.

Table 7 provides the dc conductivity values and activation energies of microcrystalline and ball-milled $Li_{1.3}Ti_{1.7}Al_{0.3}(PO_4)_{2.9}(VO_4)_{0.1}$ materials. Grain-boundary conductivity of 55h ball-milled material at 65°C illustrates an order of magnitude increase in comparison to the microcrystalline counterpart. High frequency investigation is needed to explore the grain characteristics of the microcrystalline and 22h ball-milled material. Micro-strain induced by the milling creates defects like grain-boundaries and its volume fraction is much more in ball-milled samples. Ions can diffuse faster through grain-boundaries and it is reflected in the observed jump in the conductivity in the 55h ball-milled material (Lakshmi et al, 2009, 2011).The ease of ion diffusion through grain-boundary is reflected in the values of activation energy as given in Table 7. With the milling duration activation energy decreases since the ion diffusion become easier as the volume fraction of the grain-boundaries increases.

$LATPV_{0.1}$	Grain-boundary		Grain	
	Conductivity at 65°C, σ_{dcgb} [Scm^{-1}]	Activation energy $E_{\sigma gb}$ [eV]	Conductivity at 65°C, σ_{dcg} [Scm^{-1}]	Activation energy $E_{\sigma g}$ [eV]
Microcrystalline	$3.75x10^{-8}$	(0.73±0.090)	---	----
22h ball-milled	$1.28x10^{-7}$	(0.65±0.007)	---	----
55h ball-milled	$3.13x10^{-7}$	(0.26±0.040)	$5.32x10^{-5}$	(0.30±0.01)

Table 7. Conductivity and activation energy of the grain-boundary and grain conduction in microcrystalline and ball-milled $Li_{1.3}Ti_{1.7}Al_{0.3}(PO_4)_{2.9}(VO_4)_{0.1}$ materials.

The spectroscopic plot of real part of the complex permittivity, $\varepsilon^*(\omega)$ of $Li_{1.3}Ti_{1.7}Al_{0.3}(PO_4)_{2.9}(VO_4)_{0.1}$ shows relaxation at the high frequency. This results from the constriction effect at the grain-boundaries (Mouahid et al., 2001) and is explicit in the impedance representation. This relaxation is prominent in the samples milled for longer times since the grain-boundaries are more significant in those samples (Martin et al., 2006). The ε' of $Li_{1.3}Ti_{1.7}Al_{0.3}(PO_4)_{2.9}(VO_4)_{0.1}$ material shows a prominent increase at low frequency which is associated with charges accumulating at the blocking electrode. Permittivity loss in the 55h ball-milled material shows an order of magnitude increase in comparison to the microcrystalline material and the augmented permittivity loss may be due to the ease of diffusion through the grain-boundaries that is reflected in the total conductivity hike of the 55h ball-milled material.

The complex impedance plane plots of $Na_3Cr_2(PO_4)_3$-G1:3 at 373K and 323K are given in Fig. 17(a). The equivalent circuit, $(R_gQ_g)(R_eQ_e)$, at 373K consists of a depressed semi-circle and part of a semi-circle. The impedance plane plots are depressed due to the distribution of relaxation times; a non-ideal capacitor or constant phase element, Q, is used to explain the depression (Barsoukov & Macdonald, 2005; Mariappan & Govindaraj, 2005). The high frequency part, (R_gQ_g), corresponds to grain contribution and the part of a semi-circle, (R_eQ_e) in the low frequency represents the electrode polarization [32]. Exponent $n_g=(0.93\pm0.01)$, $R_e=(3.92\pm0.42)x10^5\Omega$, $Q_e=(6.34\pm0.82)x10^{-7}S.s^n$ and $n_e=(0.69\pm0.02)$. The magnitude of chi-square is found to be $9.02x10^{-3}$. The magnitude of Q_g confirms that the high frequency contribution is from grain and not from the grain-boundary.

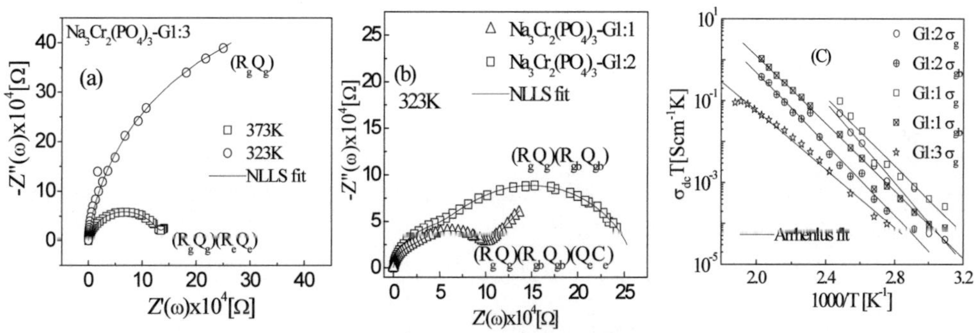

Fig. 17. (a) Complex impedance plane plot for $Na_3Cr_2(PO_4)_3$-G1:3 at 373K and 323K and the solid line represents NLLS fit to equivalent circuit (b) Complex impedance plane plot for G1:1 and G1:2 molar ratios at 323K and the solid line represents NLLS fit (c) Arrhenius plot of dc conductivity values, σ_{dcg} and σ_{dcgb}, of the three fuel molar ratios.

The $Na_3Cr_2(PO_4)_3$ with other glycine molar ratios contain contributions from both grain and grain-boundary, as evident from the two semi-circles in the complex impedance plane representation. The equivalent circuit representation of $Na_3Cr_2(PO_4)_3$-G1:1 is $(R_gQ_g)(R_{gb}Q_{gb})(Q_eC_e)$, at 323K, where, $R_g=(2.30\pm0.06)x10^4\Omega$, $Q_g=(3.46\pm0.24)x10^{-11}S.s^n$, $n_g=(0.95\pm0.06)$, $R_{gb}=(7.69\pm0.18)x10^4\Omega$, $Q_{gb}=(1.36\pm0.20)x10^{-10}S.s^n$, $n_{gb}=(0.91\pm0.02)$, $Q_e=(1.81\pm0.50)x10^{-7}S.s^n$ and $n_e=(0.46\pm0.06)$. The magnitude of chi-square is found to be $9.52x10^{-3}$. The equivalent circuit representation of $Na_3Cr_2(PO_4)_3$-G1:2 is $(R_gQ_g)(R_{gb}Q_{gb})$, at

323K where, R_g=(2.16±0.05)x10^4Ω, Q_g=(3.39±0.10)x10^{-11}S.s^n, n_g=(0.95±0.05), R_{gb}=(1.12±0.53)x10^5Ω, Q_{gb}=(5.26±0.18)x10^{-10}S.s^n and n_{gb}=(0.82±0.03). The magnitude of chi-square is found to be 9.78x10^{-3}. In these cases, grain and grain-boundary contributions are distinguished by the magnitude of constant phase element; for grain contribution, Q_g, value is in the range 10^{-12}-10^{-11}S.s^n, for grain-boundary, Q_{gb}, value is around 10^{-10}-10^{-9}S.s^n. For electrode contribution, Q_e, takes the value in the range of 10^{-7}-10^{-6}S.s^n.

The R_g and R_{gb} values are obtained by intercept of high frequency and low frequency semi-circles with the real axis and are used to calculate the dc conductivity values, σ_{dcg} and σ_{dcgb} using the cell constant. The parameters, σ_{dcg} and σ_{dcgb} are thermally activated and show Arrhenius dependence on temperature. The dc conductivity values and the activation energy, obtained from the slope of Arrhenius plot, are given in Table 8. Complex impedance plane plots for 1:1 and 1:2 glycine fuel molar ratios at 323K are shown in Fig. 17(b) and the solid line represents NLLS fit to the equivalent circuit. The highest dc conductivity value, (2.35±0.25)x10^{-6}Scm^{-1} at 323K, is obtained for $Na_3Cr_2(PO_4)_3$-G1:1 among the different glycine molar ratios. This magnitude is one order higher than the reported value, 1.1x10^{-7}Scm^{-1}, for conventionally synthesized $Na_3Cr_2(PO_4)_3$ (d'Yvoire et al.,1983.) The increase in the conductivity of $Na_3Cr_2(PO_4)_3$-G1:1 is explained through its dense sintering (Lakshmi et al., 2011,2011) (93.25% of theoretical density) and the smallest crystallite size, (31.29±3.91)nm, among the series (Lakshmi et al., 2011,2011). The present study evidenced that the grain and grain-boundary conductivity values decreases with fuel/complexing agent ratio in glycine assisted synthesis. Arrhenius plot of dc conductivity values, σ_{dcg} and σ_{dcgb}, for three glycine molar ratios are shown in Fig. 17(c). Agglomeration increases with fuel molar ratio, due to hike in the flame temperature. Agglomeration decreases the density owing to less packing of larger crystallites, which affects the electrical properties adversely. This study concluded that, the fuel molar ratio play a major role in deciding the physical and electrical properties and 1:1 fuel molar ratio is found to be the optimized value to obtain the highest electrical conductivity.

$Na_3Cr_2(PO_4)_3$	σ_{dc} [Scm^{-1}]		Activation energy [eV]		
			Conduction		Relaxation
	Grain	Grain-boundary	Grain	Grain-boundary	Grain
G1:1#	(2.35±0.25)x10^{-6}	(5.57±0.69)x10^{-7}	(0.82±0.07)	(0.81±0.02)	(0.69±0.02)
G1:2#	(2.13±0.25)x10^{-6}	(2.10±0.32)x10^{-7}	(0.97±0.08)	(0.87±0.03)	(0.72±0.01)
G1:3#	(1.75±0.15)x10^{-7}	-----	(0.71±0.02)	-----	(0.70±0.01)
U1:1*	(8.06±0.15)x10^{-7}	(2.95±0.10)x10^{-7}	(1.12±0.06)	(0.85±0.06)	(0.67±0.02)
U1:2*	(2.79±0.23)x10^{-6}	(1.29±0.24)x10^{-6}	(0.92±0.03)	(0.73±0.02)	(0.59±0.02)

*at 80°C and # at 50°C

Table 8. The dc conductivity values and activation energy of $Na_3Cr_2(PO_4)_3$ synthesized using different fuels/complexing agents.

Fig. 18(a) shows the complex impedance plot of $Na_3Cr_2(PO_4)_3$-U1:1 and $Na_3Cr_2(PO_4)_3$-U1:2 at 383K. In urea assisted $Na_3Cr_2(PO_4)_3$ series, 1:2 molar ratio showed improved conductivity due to less activation energy compared to 1:1 molar ratio, as shown in Fig. 18(b). Among the different fuels used, $Na_3Cr_2(PO_4)_3$-U1:2 showed the highest conductivity due to lower grain-boundary activation energy of (0.73±0.02)eV. The volume fraction of grain-boundary is more in nanocrystalline materials and it enhances the diffusion of ions. Table 8 gives dc conductivity and activation energy values of $Na_3Cr_2(PO_4)_3$ synthesized using different fuels.

The rhombohedral symmetry of combustion synthesised $Na_3Cr_2(PO_4)_3$ is a disordered phase. It shows higher conductivity compared to the conventionally synthesised material, may be due to the enhanced mobility owing to increase in unit cell volume compared to the microcrystalline material.

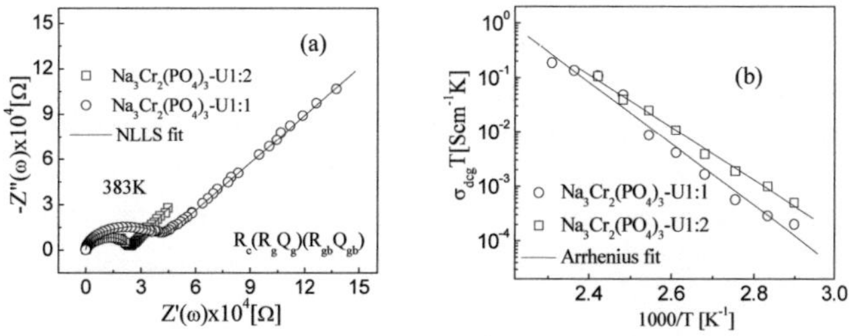

Fig. 18. (a) Complex impedance plane plot at 383K for $Na_3Cr_2(PO_4)_3$-U1:1 and $Na_3Cr_2(PO_4)_3$-U1:2. The solid line represents NLLS fit to equivalent circuit $R_c(R_gQ_g)(R_{gb}Q_{gb})$ (b) Arrhenius plots of grain dc conductivity values of $Na_3Cr_2(PO_4)_3$-U1:1 and $Na_3Cr_2(PO_4)_3$-U1:2.

The characteristic frequency of electrical relaxation in grain is obtained from the maximum of imaginary part of electric modulus or impedance spectrum. Characteristic relaxation frequencies (ω_R) obtained from Z''(ω) curve shift towards high frequency with increase in temperature. Figs. 19(a) and (b) show the spectroscopic plot of imaginary part of impedance. The $Na_3Cr_2(PO_4)_3$-G1:1 contains both grain and grain-boundary contributions at high temperatures, while $Na_3Cr_2(PO_4)_3$-G1:3 contains only grain contribution. The characteristic relaxation frequencies for grain are obtained by NLLS fitting of Z''(ω) plot. Relaxation frequency exponentially increases with temperature and its activation energy, E_h, is obtained from the Arrhenius plot, as shown in Fig. 19(c). The activation energy for electrical relaxation is given in Table 8 for $Na_3Cr_2(PO_4)_3$ material synthesized using different fuels. Such ion transport peculiarities are dominant in compounds with lithium or sodium as well as in oxygen solid electrolytes. The hopping polarization loss is responsible for the peak in the dispersive plot of Z''(ω) (Losila et al., 1998; Elliot, 1994)]. This illustrates that, while relaxing ions have to overcome less energy barrier compared to the conduction process.

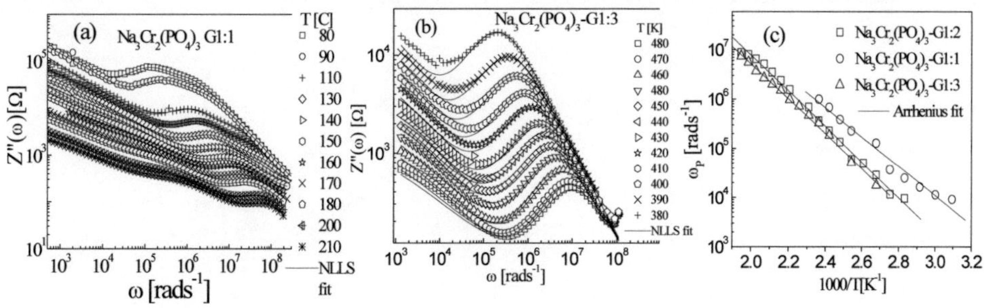

Fig. 19. Dispersion of Z''(ω) at different temperatures of (a) $Na_3Cr_2(PO_4)_3$-G1:1 (b) $Na_3Cr_2(PO_4)_3$-G1:3 and (c) Arrhenius plot of dispersion peak frequency (ω_p) of $Na_3Cr_2(PO_4)_3$-G1:1, $Na_3Cr_2(PO_4)_3$-G1:2 and $Na_3Cr_2(PO_4)_3$-G1:3.

The $Li_3Fe_2(PO_4)_3$ is synthesized using different fuels *i.e.,* glycine ($Li_3Fe_2(PO_4)_3$-G) in 1:2 molar ratio and citric acid: ethylene glycol mixture in 1:1 molar ratio ($Li_3Fe_2(PO_4)_3$-CA:EG). The complex impedance spectra of $Li_3Fe_2(PO_4)_3$-CA:EG at 373K is shown in Fig. 20(a). The equivalent circuit consists of a depressed semi-circle, $R_c(R_gQ_g)(Q_e)$, corresponding to the grain contribution and the spike, (Q_e), represents the electrode polarization at the low frequency region. R_c is the resistance of electrolyte–electrode contact, that is characterized by the shift of impedance arc from the origin. Grain-boundary contribution is observed at higher temperatures in addition to the grain contribution in this material. Fig. 20(b), shows Arrhenius temperature dependence of dc conductivity and hopping frequency of $Li_3Fe_2(PO_4)_3$ materials. $Li_3Fe_2(PO_4)_3$-G shows higher conductivity values and its dc conductivity value at 323K is $(1.14 \pm 0.05) \times 10^{-7} Scm^{-1}$. This value is around one order higher than the reported value, $5.6 \times 10^{-8} Scm^{-1}$, for the microcrystalline $Li_3Fe_2(PO_4)_3$.

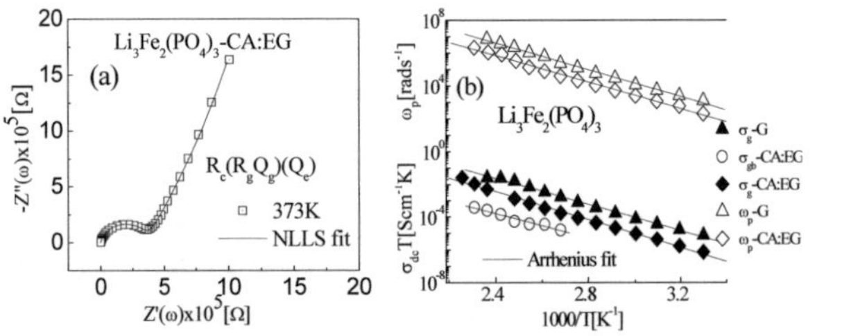

Fig. 20. (a) Complex impedance plane plot of $Li_3Fe_2(PO_4)_3$-CA:EG at 373K (b) Arrhenius plots of grain and grain-boundary dc conductivity and dispersion peak frequency.

The dc conductivity in ion conducting materials mainly depends on charge carrier density and mobility; but the carrier density is almost same for both $Li_3Fe_2(PO_4)_3$ materials (CA:EG and G) as shown in Fig. 21(a). The hopping rate and unit cell volume of $Li_3Fe_2(PO_4)_3$-G is higher than $Li_3Fe_2(PO_4)_3$-CA:EG. The improved conduction in $Li_3Fe_2(PO_4)_3$-G is resulted from the enhanced mobility and the frame-work volume (Miyajima et al., 1996). The spectroscopic plot of $Z''(\omega)$ for $Li_3Fe_2(PO_4)_3$-G is shown in Fig. 21(b). The relaxation frequencies show Arrhenius dependence on temperature and its activation energy, E_h, for investigated samples are given in Table 9. The activation energy for electrical relaxation is almost same for both $Li_3Fe_2(PO_4)_3$ prepared using different fuels/complexing agents. While relaxing, ions have to overcome less energy barrier compared to the conduction process.

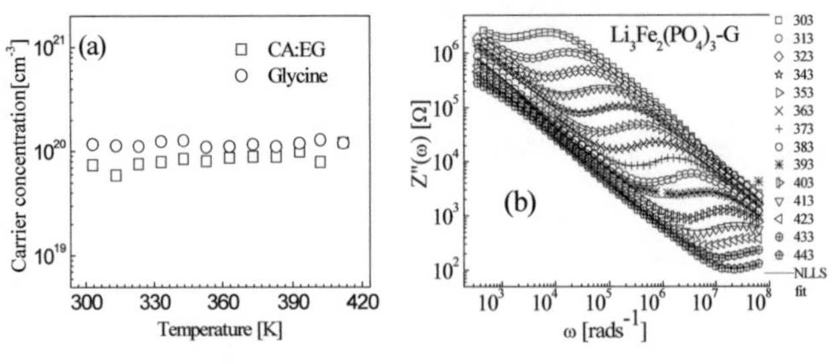

Fig. 21. (a) Chrage carrier concentration, n_c, of $Li_3Fe_2(PO_4)_3$ over the range of temperature 300K-420K and (b) Frequency dependence of imaginary part of impedance of $Li_3Fe_2(PO_4)_3$-G showing grain contribution to relaxation.

Complex impedance plot of $Na_3Fe_2(PO_4)_3$ contains depressed semi-circle and the low frequency electrode effect is as shown in Fig. 22(a). The circuit description is $R_c(R_gQ_g)(Q_eC_e)$, where R_c is the contact resistance, that is characterized by the shift of impedance arc from the origin. The (R_gQ_g) and (Q_eC_e) correspond to grain and electrode contribution. Typical value of Q_e and Q_g are of the order of 10^{-7} and $10^{-12}S.s^n$ respectively. Arrhenius plot of grain dc conductivity is shown in Fig. 22(b) and the activation energies for conduction and relaxation are given in Table 9. The dc conductivity value of $Na_3Fe_2(PO_4)_3$ is higher than $Li_3Fe_2(PO_4)_3$ due to the rattling of Li^+ ions within the large interstitial space available (Shannon et al. ,1977). $Na_3Fe_2(PO_4)_3$ synthesized by the present technique show one order increase in conductivity compared to the conventional microcrystalline material. Solution combustion synthesized $Li_3Fe_2(PO_4)_3$ and $Na_3Fe_2(PO_4)_3$ are crystallized in monoclinic symmetry *i.e.,* β-$Fe_2(SO_4)_3$ type structure. In this structure, mobile ion occupies a four co-ordination site in contradiction with the NASICON structure, where it occupies six co-ordination sites. The four co-ordination site of monoclinic structure is preferable to the six co-ordination site of NASICON for ion conduction (Nomura et al., 1993).

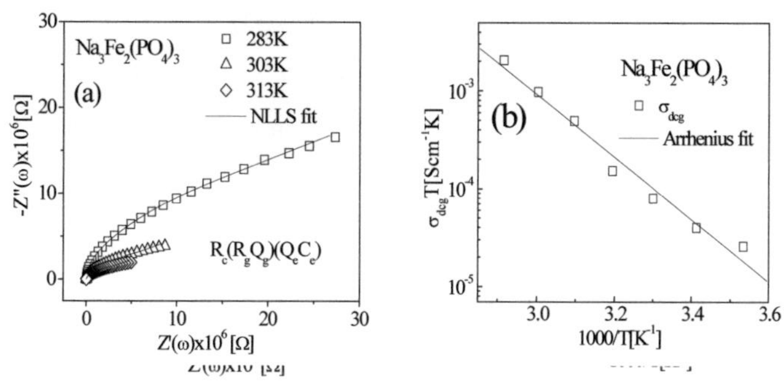

Fig. 22. (a) Complex impedance plane plot of $Na_3Fe_2(PO_4)_3$ at different temperatures (b) Arrhenius plot of grain dc conductivity.

Material	σ_{dcg} at 323K [Scm^{-1}]	Activation energy [eV]	
		Conduction	Relaxation, ω_p
$Li_3Fe_2(PO_4)_3$-CA:EG	$(1.34\pm0.01)x10^{-8}$	(0.88 ± 0.03)	(0.81 ± 0.02)
$Li_3Fe_2(PO_4)_3$-Glycine	$(1.14\pm0.05)x10^{-7}$	(0.79 ± 0.02)	(0.81 ± 0.01)
$Na_3Fe_2(PO_4)_3$	$(1.52\pm0.03)x10^{-6}$	(0.63 ± 0.04)	----

Table 9. The grain dc conductivity values at 323K and grain activation energies for dc conduction and electrical relaxation of $Li_3Fe_2(PO_4)_3$ and $Na_3Fe_2(PO_4)_3$.

4.1 Modulus representation and scaling analysis

Macedo *et al.,* (Macedo et al., 1972 & Moynihan et al., 1974) formulated a theory for conductivity relaxation in ion conductors in terms of a dimensionless quantity, complex electric modulus, $M^*(\omega)$. In modulus formalism the details at low frequencies are suppressed. Imaginary part of the complex electrical modulus in the frequency domain is owing to Kohlrausch William Watts relaxation function has been found to be well approximated by Bergman. By the fitting of $M''(\omega)$, the parameters like M''_p, β and ω_p are extracted. The maximum value of M_p'' is found at $\omega_p=1/\tau_p$ and ω_p shifts to higher frequencies with increase in temperature.

The charge carriers are mobile over long distances in the region left to peak; while right to the peak ions are spatially confined to the potential wells. The frequency of relaxation, ω_p, where $M_p''(\omega)$ is an indicative of transition from short-range to long-range mobility at the decreasing frequency. The ω_p exponentially increases with temperature and the activation energy for relaxation is calculated from the Arrhenius behaviour. The scaling of modulus spectra is shown in Fig. 23(a), for $Na_3Cr_2(PO_4)_3$-G1:3 and inset shows the Arrhenius plot of ω_p. Grain contribution is dominant in $Na_3Cr_2(PO_4)_3$-G1:3, over the frequency and temperature range of the experiment.

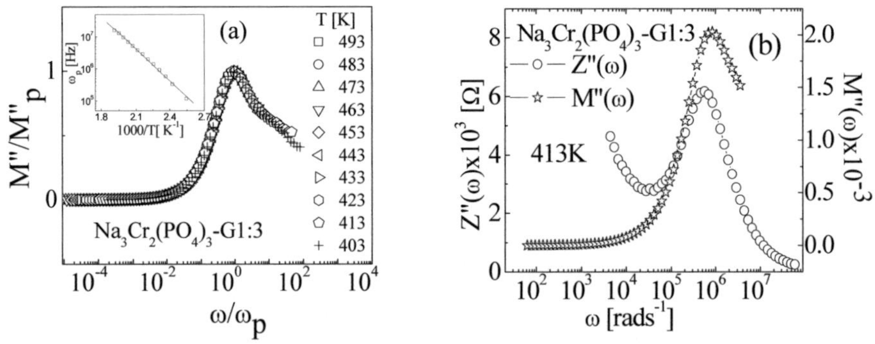

Fig. 23. (a) The modulus scaling in $Na_3Cr_2(PO_4)_3$-G1:3 at different temperatures and (b) Frequency dependence of $Z''(\omega)$ and $M''(\omega)$ at 413K.

Further, in Fig. 23(b), $Z''(\omega)$ and $M''(\omega)$ peaks are almost coincident and there is no additional peak in these representations. The single relaxation peak in the modulus representation of $Na_3Cr_2(PO_4)_3$-G1:3 is contributed from the grain part since the electrode contribution is suppressed. The $Z''(\omega)$ and $M''(\omega)$ peaks are almost coincident, which implies that the grain contributes for impedance relaxation. The small separation in the modulus and impedance peak positions points to the good grain connectivity.

Material	Activation energy for relaxation, E_h[eV]
$Na_3Cr_2(PO_4)_3$-G1:3	(0.66±0.01)
$Na_3Cr_2(PO_4)_3$-U1:1	(0.93±0.04)
$Na_3Cr_2(PO_4)_3$-U1:2	(0.89±0.04)
$Li_3Fe_2(PO_4)_3$-CA:EG	(0.71±0.02)

Table 10. The activation energy for electrical relaxation obtained from Modulus representation

Conductivity spectra at different temperatures collapsed to a single curve at higher frequencies on appropriate scaling, which implies that the relaxation mechanism at the higher frequency is independent of temperature. But in some cases, as shown in Figs. 24(a)-(b), low frequency part of the plot is not scaled due to contribution from electrode polarization. The present study explored Ghosh ($\sigma'(\omega)/\sigma_{dc}=F(\omega/\sigma_{dc}T)$) and Summerfield methods ($\sigma'(\omega)/\sigma_{dc}=F(\omega T_s)$) for scaling analysis. The $Na_3Cr_2(PO_4)_3$-G1:3 scaled better for Summerfield method than Ghosh's formalism (Ghosh & Pan, 2000; Summerfield, 1985), since it uses directly available parameters for scaling. Thus, scaling of conductivity and modulus spectra provided the time-temperature superposition principle of ion dynamics in the material.

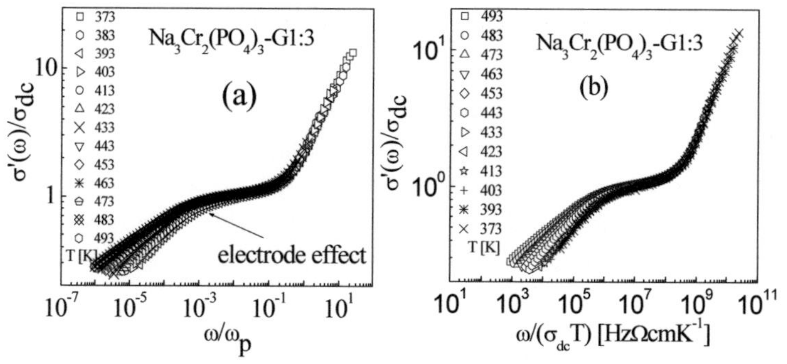

Fig. 24. The conductivity scaling of $Na_3Cr_2(PO_4)_3$-G1:3 using (a) Ghosh formalism and (b) Summerfield formalism.

5. Conclusion

In the present study, NASICON materials of two different symmetry, *i.e. rhombohedral (NASICON type) and modifications of monoclinic ($Fe_2(SO_4)_3$-type)*, are investigated. Different characterization techniques are used for the confirmation of structural, magnetic and electrical properties. The main initiative of the present study is to correlate the ion mobility with the symmetry.

Out of these, $LiTi_2(PO_4)_3$ family based on rhombohedral symmetry is synthesized by high energy ball-milling. Due to strain effect, defects like grain-boundaries are introduced in these materials. These grain-boundaries are less activation energy path for mobile ions and thus enhancing the electrical conductivity. The $A_3M_2(PO_4)_3$ (where A=Li, Na and M=Fe, Cr) family is prepared by solution combustion technique. By solution combustion synthesis technique, thermal stability is achieved for room temperature phase of $Na_3Cr_2(PO_4)_3$ and $Li_3Fe_2(PO_4)_3$ materials. The fuels/complexing agents played a major role in controlling the physical and electrical properties in these materials. This study concluded that, the fuel molar ratio play a major role in deciding the physical and electrical properties and 1:1 glycine molar ratio is found to be the optimized value to obtain the highest electrical conductivity in $Na_3Cr_2(PO_4)_3$ materials. While, the charge carrier density in $Na_3Cr_2(PO_4)_3$ and $Li_3Fe_2(PO_4)_3$ was independent of the fuels/complexing agents.

Structural distortions, involving a symmetry lowering to orthorhombic, monoclinic or triclinic, are possible and that may affect the disorder state and mobility of lithium/sodium substantially. Mobile cation occupies a six coordination site in the NASICON-type structure and a four coordination site in the $Fe_2(SO_4)_3$-type compounds. The activation energies for ionic conduction of $Fe_2(SO_4)_3$-type structure is a little lower than that of the NASICON. This indicates that the four coordination site of the $Fe_2(SO_4)_3$-type structure is preferable to the six coordination of the NASICON-type structure for ionic conduction. This is the reason behind the enhanced conduction in combustion synthesized $Li_3Fe_2(PO_4)_3$ and $Na_3Fe_2(PO_4)_3$ materials.

The scaling of ac conductivity and modulus spectra provided time-temperature superposition principle of ion dynamics in these materials. The ability to scale different data sets to one common curve indicated that the common physical mechanism in a process can be separated by thermodynamic scales. These materials find application in sensors, rechargeable batteries *etc.*

6. Acknowledgment

I would like to thank UGC-SAP1 F.530/15/DRS/2009 for financial support and Dr. G. Saini for TEM measurement. Central Instrumentation facility, Pondicherry University, is gratefully acknowledged for TG-DTA, FT-IR, SEM, VSM and UV-vis techniques.

7. References

Alamo, J. Roy, R. (1986). *J. Mater. Sci.*, Vol. 21, pp. 444.

Aono, H. Sugimoto, E. Sadoka, Y. Imanaka, N. Ya, G. Adachi (1993). *J. Electrochem. Soc.*, Vol.140, pp. 1827

Barsoukov, E. Macdonald, J. R. (2005). *Impedance Spectroscopy Theory, Experiment and Applications*, (Second Ed.), A John Wiley & Sons, New Jersey

Bates, T. Mackenzie, J. D. (1962). *Modern Aspects of the Vitreous State,* Vol. 2, Butterworths, London, p. 195;

Benmokhtar, S. El Jazouli, A. Krimi, S. Chaminade, J. P. Gravereau, P. Menetrier, M. De Waal, D. (2007). *Materials Research Bulletin*, Vol. 42, pp. 892.

Boukamp, B. A. (1989). *Equivalent Circuit, Users Manual,* p. 12, University of Twente, Enschede, Netherlands.

Bykov, A. B. Chirkin A. P., Demyanets, L. N. Doronin, S. N. Genkina, E. A. Ivanov-Shits, A. K. Kondratyuk, I. P. Maksimov, B. A. Melonikov, O. K. Muradyan, L. N. Simonov, V. I. Timofeeva, V. A. (1990). *Solid State Ionics, Vol.* 38, pp. 31

Corbridge D. E. C., Lowe, E. J. (1954). *J. Chem. Soc.*, Part I, pp. 493.

Delshad Chermahini, M. Sharafi, S. Shokrollahi H., Zandrahimi, M. (2009) *J. Alloys and Compounds,* Vol. 474, pp. 18

d'Yvoire, F. Pintard-ScrGpell, M. Bretey, E. de la Rochsre, M. (1983) *Solid State Ionics, Vol.* 9-10, pp.851

Edwards, J. S. Paul, A. Douglas, R. W. (1972). *Phys. Chem. Glasses, Vol.* 13, pp. 131.

ElBatal, H.A. Ezz Eldin, F. M. Shafi, N. A. (1988). *Phys.Chem. Glasses*, Vol. 29, pp. 235

Elliott, S. R. (1994). *Solid State Ionics,* Vol. 70–71, pp. 27

Ghosh, A. Pan A., (2000). *Phys. Rev. Lett.*, Vol. 84, pp. 2188

Hahn, H. Logas, J. Averback, R. S. (1990). *J. Mater. Res., Vol.* 5, pp. 609.

Hamzaoui, R., Elkedim, O. Fenineche, N., Gaffet, E. Craven, (2003). *J. Mater. Sci. Eng. A,* Vol. 360, pp. 299.

Hong, H.Y-P. (1976). *Mater. Res. Bull.*, Vol.11, pp. 173

Kravchenko, V. V. Michailov, V. I. Segaryov, S. E. (1992). *Solid State Ionics*, Vol. 50, pp. 19

Kurkjian, C. R. Sigety, E. A. (1968). *Phys. Chem. Glasses*, Vol. 9, pp. 73
Lakshmi Vijayan, Govindaraj, G. (2009) *Physica B*, Vol. 404, pp. 3539
Lakshmi Vijayan, Govindaraj, G. (2011) *J. Phys. Chem. Solids, Vol.* 72, pp. 613.
Lakshmi Vijayan, Rajesh Cheruku, Govindaraj, G. Rajagopan, S. (2011). *Materials Chemistry and Physics, Vol.* 125, pp. 184.
Lakshmi Vijayan, Rajesh Cheruku, Govindaraj, G. Rajagopan, S. (2011). *Materials Chemistry and Physics,* in press.
Larson, A. C. Von Dreele, R. B. (1994). *Los Alamos National Laboratory Report,* pp. 86.
Li, J.Q. Matsui, Y. Park, S.K. Tokura, Y. (1997). *Phys. Rev. Lett.* Vol. 79, pp. 297
Losila, E. R. Aranda, M. A. G. Bruque, S. Paris, M. Sanz, J. West, A. R. (1998). *Chem. Mater. Vol.* 10, pp. 665
Macdonald, J. R. (1987). *Impedance Spectroscopy*, Wiley, New York.
Macedo, P. B. Moynihan, C.T. Bose, R. (1972). *Phys. Chem. Glass.*, Vol. 13, pp. 171
Mariappan, C. R. Govindaraj, G. (2004). *Physica B*, Vol. 353, pp. 65
Mariappan, C. R. Govindaraj, G. (2006). *J. Non-Cryst. Solids, Vol.* 352, pp. 2737
Mariappan, C. R. Govindaraj, G. Roling, B. (2005). *Solid State Ionics*, Vol. 176, pp. 723
Martin G. Bellino, Diego G. Lamas, Noemí E. Walsoe de Reca, (2006). *Adv. Mater.* Vol. 18, pp. 3005.
Mineo Sato. Shigehisa Tajimi. Hirokazu Okawa. Kazuyoshi Uematsu. Kenji Toda. (2002). *Solid State Ionics, Vol.* 152-153, pp. 247.
Miyajima, Y. Saito, Y. Matsuoka, M. Yamamoto, Y. (1996). *Solid State Ionics*, Vol. 84, pp. 61.
Mouahid F. E., Zahir, M. Maldonado-Manso, P. Bruque, S. Losilla, E. R. Aranda, M. A. G. Leon, C. Santamaria, J. (2001). *J. Mater. Chem.* Vol. 113, pp. 258.
Moynihan, C. T. Boesch, L. P. Laberge, N. L. (1973). *Phys. Chem. Glass.* Vol. 14, pp. 122.Nobuya Machida. Hidekazu Yamamoto. Seiji Asano. Toshihiko Shigematsu. (2005).
Nomura, K. Ikeda, S. Ito, K. Einaga, H. (1993). *Solid State Ionics*, Vol. 61, pp. 293.
Palani Balaya, Martin Ahrens, Lorentz Kienle, Joachim Maier, (2006). *J. Am. Ceram. Soc.* Vol. 89, pp. 2804.
Prithu Sharma. Krishanu Biswas. Amit Kumar Mondal. Kamanio Chattopadhyay. (2009) *Scripta Materialia*, Vol. 61, pp. 600.
Puclin, T. Kaczmarek, W. A. Ninham, B. W. (1995). *Mat. Chem. Phys.*, Vol. 40, pp. 73.
Rao, K. J. Sobha, K. C. Kumar, S. (2001). *Proc. Indian Acad. Sci.*, Vol. 113, pp. 497.
Rougier, C. J. Nazri, G. A. Julian, C. (1997). *Mater. Res. Soc. Symp. Proc.*, Vol. 453, pp. 647.
Rulmont, A. Cahay, R. Liégeois-Duyckaerts, M. Tarte, P. (1991). *Eur. J. Solid State Inorg. Chem., Vol.* 28, pp. 207.
Savosta, M. M. Krivoruchko, V. N. Danilenko, J. A. Yu. Tarenkov, V. Konstantinova, T. E. Borodin, A. V. Varyukhin, V. N. (2004). *Phys. Rev. B*, Vol. 69, pp. 24413.
Sayer, M. Mansingh, A. (1972). *Phys. Rev. B,* Vol. 6, pp. 4629.
Schoonman, J. (2003). *Solid State Ionics, Vol.* 157, pp. 319.
Shannon, R. D. Taylor, B. E. English, A. D. Berzins, T. (1977) *Electrochim. Acta*, Vol. 22, pp. 783
Sigaryov, S. E. (1992). *Mater. Sci. Eng. B*, Vol. 13, pp. 121.
Stalhandske, C. (2000). Glasteknisk Tidskrift, Vol. 55, pp. 65.
Steele, F. N. Douglas, R. W. (1965). *Phys. Chem. Glasses, Vol.* 6, pp. 246
Summerfield, S. (1985). *Philos. Mag. B, Vol.* 52, pp. 9.

Takano, M. Kawachi, J. Nakanishi, N. Takeda, Y. (1981). *J. Solid State Chem.*, Vol. 39, pp. 75

Toby, B. H. (2001). *J. Appl. Cryst.*, Vol. 34, pp.210

Unit-Cell software for cell refinement method of TJB Holland & SAT Redfern, 1995.

West, A. R. Abram, E. J. Sinclair, D. C. (2002). in: *Proceeding of 8th Asian conference on Solid State Ionics*, p.487, World Scientific, Singapore

Williamson, G. K. Hall, W. H. (1953). *Acta. Metall.*, Vol.1, pp. 22.

Wong Shan, Peter J. Newman, Best, A. S. Nairn, K.M. Macfarlane, D. R. Maria Forsyth (1998). *J. Mat. Chem.*, Vol. 8, pp. 219.

Yamamoto Hidekazu. Machida Nobuya. Shigematsu Toshihiko. (2004). *Solid State Ionics*, Vol. 175, pp. 707

5

Structural Characterization of New Perovskites

Antonio Diego Lozano-Gorrín
Universidad de La Laguna
Spain

1. Introduction

The perovskite is a calcium titanium oxide mineral species composed of calcium titanate, with the chemical formula $CaTiO_3$. The mineral was discovered in the Ural mountains (Russia) by Gustav Rose in 1839 and is named after Russian mineralogist L. A. Perovski (1792-1856).

The perovskite crystal structure was published in 1945 from X-ray diffraction data on barium titanate by the Irish crystallographer H. D. Megaw (1907-2002). It is a true engineering ceramic material with a plethora of applications spanning energy production (SOFC technology), environmental containment and communications.

The general stoichiometry of the perovskite structure is ABX_3, where *A* and *B* are cations and X is an anion. The *A* and *B* cations can have a variety of charges and in the original perovskite mineral ($CaTiO_3$) the *A* cation is divalent and the *B* cation is tetravalent. $CaTiO_3$ exhibits an orthorhombic structure with space group Pnma.

The traditional view of the perovskite lattice is that it consists of small *B* cations within oxygen octahedra, and larger *A* cations which are XII fold coordinated by oxygen.

Figure 1 shows a picture of the mineral perovskite which, as it was mentioned above, is composed by calcium titanate.

Fig. 1. Perovskite mineral species ($CaTiO_3$).

2. Methods of synthesis

There are different methods of synthesis used in the field of Solid State Chemistry. Among them, one of the most used is that known as ceramic method. The ceramic method consists

of the grinding of the stoichiometric quantities of the corresponding starting compounds and subsequent heating by using furnaces at relatively high temperatures. However, it is not the most efficient method many times.

The sol-gel method and, used more and more every day, the freeze-drying method, seem to offer purer materials by reducing the heating time and working at not so high temperatures. Besides, other methods complete the list of synthesis processes of solid materials.

2.1 Ceramic method

The well known as ceramic method consists of the mixing and grinding of the starting compounds in stoichiometric quantities, being afterwards heated in furnaces at relatively high temperatures for relatively long dwelling times. Eventually, the desired material will be obtained, with a smaller or higher grade of impurities.

The main disadvantage of this method is the low homogeneity of the obtained product, which demands repeated procedures of intermediate homogenization and thermal treatment at high temperature. Thus, the method requires the grinding (intermittently) of the material during the heating treatment in order to minimize the drawback created by the interface (Figure 2) resulting of the reaction between the starting reagents. Figure 2 shows how that interface would be formed by A and B chemically combined. That interface increases its size as a function of the time of reaction, giving rise to a slower and less efficient transfer of A and B in opposite direction to meet and react. Thus, by grinding the grain formed by A, B, and AB, one will be able to increase the contact surface between A and B and therefore to make more efficient the reaction. We should just take into account that the solid state reactions are extremely slow, and that is why the high temperature and high dwelling times are required, but besides, a high contact surface between the reagents is desired to increase the velocity of the reaction. As a result, high crystalline materials are obtained, which are not acceptable for catalytic applications.

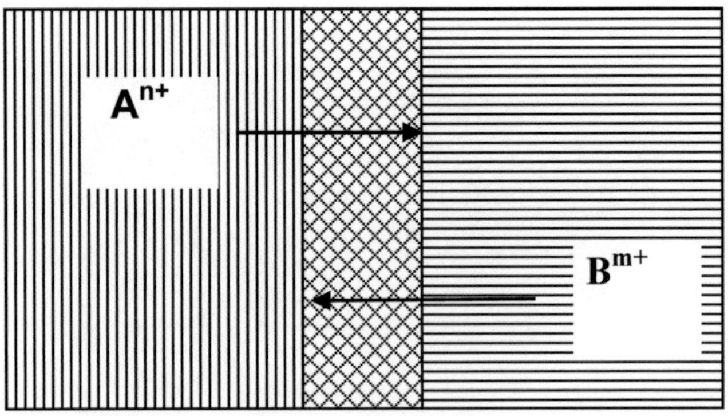

Fig. 2. Reaction between grains corresponding to A^{n+} and B^{m+}. The cross-lined area corresponds to the interface created between them.

2.2 Sol-gel method

The sol-gel method is a wet-chemical process widely used in the field of materials science. The starting point is the colloidal solution (*sol*) that acts as the precursor for an integrated network (*gel*) of either discrete particles or network polymers.

Perovskites can be synthesized starting from the oxides or salts of the different metals wanted to be part of the final structure by dissolving them in a solution containing a complexing agent (citric acid, e.g.) and heating it up to form gel. Immediately after, a drastic heating leads to the decomposition of the organic part and therefore to the formation of a fine and intimate distribution of the different metals in the resulting precursor, meaning this a high contact surface between them that allows a more efficient reaction at a relatively high temperature and a prudent dwelling time.

2.3 Freeze-drying method

The freeze-drying method, also known as lyophilisation, is a process consisting of the freezing of the solution of the starting reagents and the subsequent reducing of the surrounding pressure to allow the frozen solvent to sublime directly from the solid phase to the gas phase.

In the lab, the process is carried out by placing the solution in a decantation funnel and dropping it slowly in a container with liquid nitrogen in order to freeze it in small rounded "grains". Right after the dropping ends, the frozen 'grains' of solution are placed in an Erlenmeyer flask which is immediately connected to a freeze-dryer (Figure 3). Now the

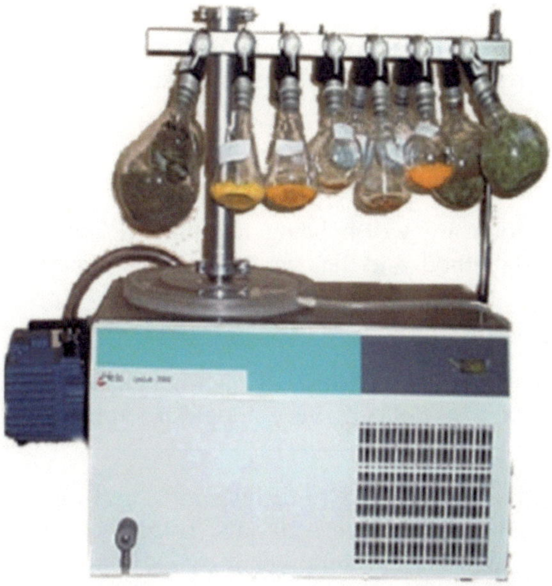

Fig. 3. Freeze-dryer with several Erlenmeyer flasks connected containing different frozen solutions.

freeze-dryer will carry out its function: to provoke the sublimation of the solvent. For that, the frozen solution undergoes an important vacuum created by a pump, allowing the solvent to sublimate and being the latter frozen once again into a big drum at a very low temperature inside the device.

After one/two day(s) the dried solution - called now *precursor* - is disconnected from the freeze-dryer and treated at certain temperature in order to allow the reaction between the different reagents which are mixed in extremely fine and close way. This method allows much lower both dwelling times and heating temperatures, at the time that makes easier the way to get pure phases.

2.4 Melt method

One method often employed is to melt the reagents together and then later anneal the solidified melt. If volatile reagents are involved these are usually put in an ampoule that is evacuated (the reagent mixture is kept cold - by keeping the bottom of the ampoule in liquid nitrogen, e.g.) and then sealed. The sealed ampoule is then put in an oven/furnace and given a certain heat treatment.

2.5 Pechini method

Pechini method is a widely used method consisting of the mix of metal nitrate solution of the starting materials with a stoichiometric amount of citric acid. The resulting solution is stirred on a hot plate and the temperature stabilized at about 90ºC, at which point ethylene glycol is added at a mass ratio of 40:60 with respect to citric acid. The temperature is maintained constant up to the resin formation, which polymerizes at about 300ºC. The precursor powders are heated for several hours at various temperatures.

This method, as a soft-chemistry one, leads to homogeneous and low crystalline products at relatively low temperatures, but they usually require expensive initial compounds and/or complicated synthesis procedures.

2.6 Other methods

Some relative new methods are worth being mentioned, like the solution combustion method, the sonochemical method, and the microwave assisted synthesis.

2.6.1 Solution combustion method

Combustion synthesis is an effective, low-cost method for production of various useful materials. Today, it has become a very popular approach for preparation of nanomaterials.

Solution combustion synthesis (SCS) is a versatile, simple and rapid process, which allows effective synthesis of variety of nanosize materials. This process involves a self-sustained reaction in homogeneous solution of different oxidizers (e.g., metal nitrates) and fuels (e.g., urea, glycine, hydrazides). Depending on the type of the precursors, as well as on conditions used for the process organization, the SCS may occur as either volume or layer-by-layer propagating combustion modes. This process not only yields nanosize oxide materials but also allows uniform (homogeneous) doping of trace amounts of rare-earth impurity ions in a single step.

2.6.2 Microwave assisted method

Microwave heating allows a rapid heating rate, however, the final yield decreases compared with the conventional methods. It leads to the synthesis of materials by consumption of less energy.

2.6.3 Sonochemical method

The sonochemical method is that in which the molecules undergo a chemical reaction due to the application of powerful ultrasound radiation (20 kHz–10 MHz).

Ultrasonic irradiation differs from traditional energy sources (such as heat, light, or ionizing radiation) in duration, pressure, and energy per molecule.

3. Perovskite-type structures

Many transition metal oxides show the very versatile perovskite structure. The rich variety of physical properties such as high-temperature superconductivity and colossal magnetoresistance observed in these compounds makes them very attractive from both fundamental and applied perspectives.

The general chemical formula for perovskite compounds is ABX_3, where *A* and *B* are two cations of very different sizes, and X is an anion that bonds to both. The *A* atoms are larger than the *B* atoms, and besides, its ionic radii close to that on the anion X, thus they can form together a cubic close packing. The smaller *B* cation is usually a 3d-transition metal ion which occupies the octahedral interstices of the close packing. X is oxygen or a halide ion. The ideal cubic-symmetry structure has the *B* cation in 6-fold coordination, surrounded by an octahedron of anions, and the *A* cation in 12-fold cuboctahedral coordination. In short, in can be also described as a network of edge sharing octahedra BX_6. The relative ion size requirements for stability of the cubic structure are quite stringent, so slight buckling and distortion can produce several lower-symmetry distorted versions, in which the coordination numbers of *A* cations, *B* cations or both are reduced. The orthorhombic and tetragonal phases are the most common non-cubic variants.

For the stoichiometric oxide perovskites, the sum of the oxidation states of *A* and *B* cations should be equal to six. The occupancy of *A* and *B* positions of different ions with appropriate ionic radii as well as for mixed occupancy of both cation positions leaded to the preparation of numerous compounds with wide spectrum of physical and chemical properties. Among the most famous representatives of perovskite class are the dielectric $BaTiO_3$, high-temperature superconductor $YBa_2Cu_3O_{7-x}$, materials exhibiting colossal magnetoresistance $R_{1-x}A_xMnO_3$, where R = La^{3+}, Pr^{3+} or other rare earth ion, A = Ca^{2+}, Sr^{2+}, Ba^{2+}, multiferroic materials, etc.

The structure of an ideal cubic perovskite is shown in Figure 4, where the *A* cations are shown at the corners of the cube, and the *B* cation in the centre with oxygen ions in the face-centred positions. The space group for cubic perovskites is Pm3m (221); the equivalent positions of the atoms are detailed in Table 1.

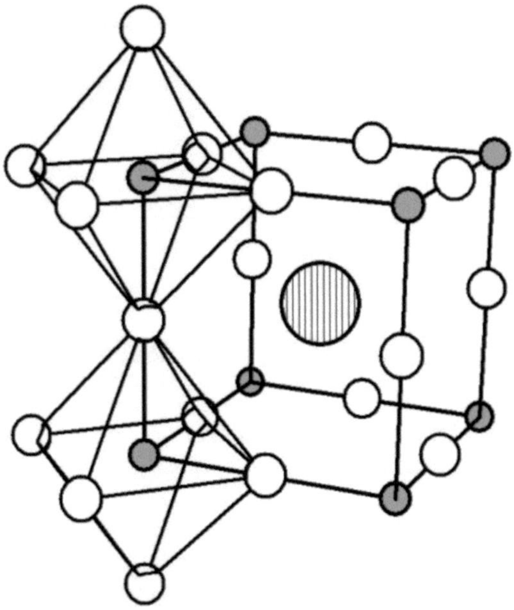

Fig. 4. Structure of the ideal cubic perovskite, ABO_3 (*A*: big stripped circle, *B*: small grey circles, O: empty circles).

Perovskite materials exhibit many interesting and intriguing properties from both the theoretical and the application point of view. Colossal magnetoresistance, ferroelectricity, superconductivity, charge ordering, spin dependent transport, high thermopower and the interplay of structural, magnetic and transport properties are commonly observed features in this family. These compounds are used as sensors and catalyst electrodes in certain types of fuel cells and are candidates for memory devices and spintronics applications.

Site	Location	Coordinates
A cation	2*a*	(0,0,0)
B cation	2*a*	(1/2,1/2,1/2)
O anion	6*b*	(1/2,1/2,0)(1/2,0,1/2)(0,1/2,1/2)

Table 1. Atomic positions in cubic perovskites

Many superconducting ceramic materials (the high temperature superconductors) have perovskite-like structures, often with 3 or more metals including copper, and some oxygen positions left vacant. One prime example is yttrium barium copper oxide which can be insulating or superconducting depending on the oxygen content.

Chemical engineers are considering this material as a replacement for platinum in catalytic converters in diesel vehicles.

Figures 4 and 5 show different ways to represent the perovskite structure. Figure 5 represents the undistorted cubic structure; the symmetry is lowered to orthorhombic, tetragonal or trigonal in many perovskites.

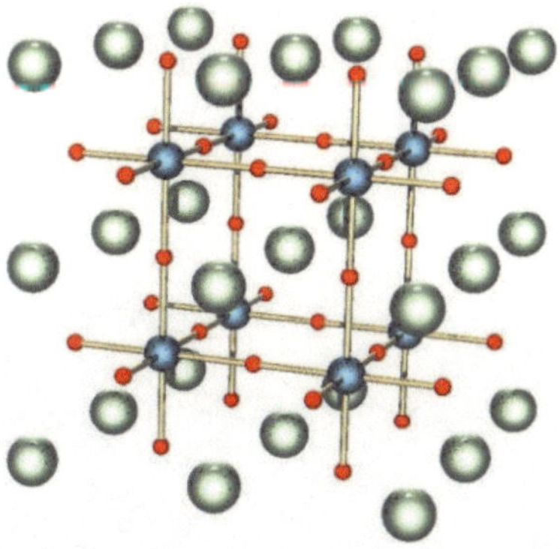

Fig. 5. Perovskite structure (ABX_3). The small spheres are X atoms (usually oxygens), the medium spheres are *B*-atoms (a smaller metal cation, such as Ti(IV)), and the big spheres are the *A*-atoms (a larger metal cation, such as Ca(II)).

4. Structural characterization of perovskites

The synthetic methodology and the characterization of the perovskite often go hand in hand in the sense that not one but a series of reaction mixtures are prepared and subjected to heat treatment. The stoichiometry is typically varied in a systematic way to find which ones will lead to new solid compounds or to solid solutions between known ones. A prime method to characterize the reaction products is X-ray powder diffraction (XRD), because many solid state reactions will produce polycristalline powders. Thus, powder diffraction will facilitate the identification of known phases in the mixture. If a pattern is found that is not known in the diffraction data libraries an attempt can be made to index the pattern, i.e. to identify the symmetry and the size of the unit cell. Obviously, if the product is not crystalline enough the characterization is typically much more difficult.

Once the unit cell of a new phase is known, the next step is to establish the stoichiometry of the phase. This can be done in a number of ways. Sometimes the composition of the original mixture will give a clue, if one finds only one product -a single powder pattern- or if one was trying to make a phase of a certain composition by analogy to known materials but this is rare. Often considerable effort in refining the synthetic methodology is required to obtain a pure sample of the new material. If it is possible to separate the product from the rest of the reaction mixture elemental analysis can be used. Another ways involves SEM and the generation of characteristic X-rays in the electron beam.

The easiest way to solve the structure is by using single crystal X-ray diffraction. The latter often requires revisiting and refining the preparative procedures and that is linked to the question which phases are stable at what composition and what stoichiometry. In other words what does the phase diagram looks like. An important tool in establishing this is thermal analysis techniques like DSC or DTA and, increasingly also, the synchrotron temperature-dependent powder diffraction. Increased knowledge of the phase relations often leads to further refinement in synthetic procedures in an iterative way. New phases are thus characterized by their melting points and their stoichiometric domains. The latter is important for the many solids that are non-stoichiometric compounds. The cell parameters obtained from XRD are particularly helpful to characterize the homogeneity ranges of the latter.

In order to analyse the different morphological and surface characteristics of particles in the perovskites, SEM (scanning electron microscopy) can be used. Figure 6 shows a micrograph obtained for $La_{0.75}Sr_{0.25}Cr_{0.2}Fe_{0.8}O_3$ (the figure shows the texture and relief created by the elimination of volatile substances produced in the combustion of organic compounds during thermal treatment).

Fig. 6. SEM for $La_{0.75}Sr_{0.25}Cr_{0.2}Fe_{0.8}O_3$.

4.1 Further characterization

In many -but certainly not all- cases new solid compounds are further characterized by a variety of techniques that straddle the fine line that (hardly) separates solid-state chemistry from solid-state physics.

4.1.1 Optical properties

For non-metallic materials it is often possible to obtain UV/VIS spectra. In the case of semiconductors that will give an idea of the band gap. Surfaces of ABO_3 perovskites and defects therein are extremely relevant for such important fields of technology as photocatalysis, gas-sensors and for applications as materials for ferroelectric memories, and the optical properties of them can be proved to be extremely interesting.

4.1.2 Electrical properties

Four-point (or five-point) probe methods are often applied either to ingots, crystals or pressed pellets to measure resistivity and the size of the Hall effect. This gives information on whether the compound is an insulator, semiconductor, semimetal or metal and upon the type of doping and the mobility in the delocalized bands (if present). Thus, important information is obtained on the chemical bonding in the material.

The impedance spectroscopy is also a very useful method used for electrical properties characterization of perovskite-type materials, especially for bulk ceramic samples.

4.1.3 Magnetic properties

Magnetic susceptibility can be measured as a function of temperature to establish whether the material is a *para-*, *ferro-*, *ferri-* or *antiferro-* magnet, among others. Again the information obtained pertains to the bonding in the material. This is particularly important for transition metal compounds. In the case of magnetic order neutron diffraction can be used to determine the magnetic structure.

Magnetic measurements are usually carried out with a SQUID magnetometer under different applied magnetic field. Figure 7 shows a picture of one of those SQUID magnetometers.

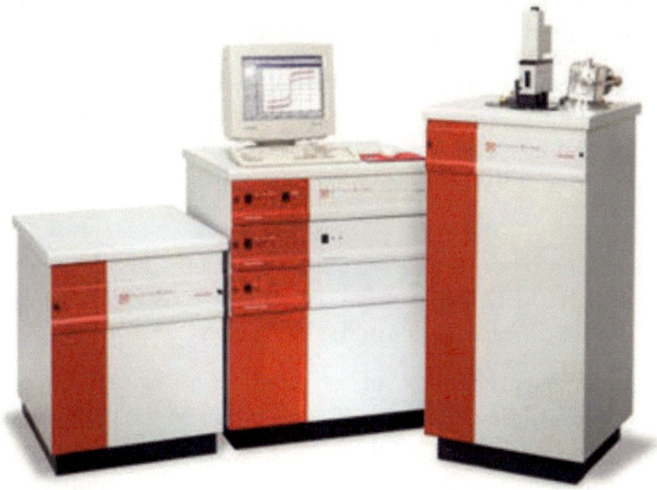

Fig. 7. SQUID magnetometer.

As an example, Figure 8 shows the magnetic susceptibility versus temperature data for the perovskite $Sr_2Ru_{0.5}Co_{1.5}O_6$.

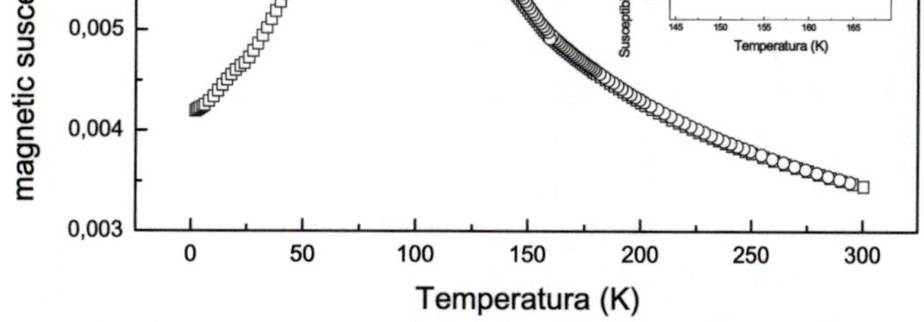

Fig. 8. Magnetic susceptibility versus temperature for $Sr_2Ru_{0.5}Co_{1.5}O_6$.

5. X-ray diffraction

About 95% of all solid materials can be described as crystalline. In 1919 A. W. Hull published a paper titled "A New Method of Chemical Analysis". Here he pointed out that "... every crystalline substance gives a pattern; the same substance always gives the same pattern; and in a mixture of substances each produces its pattern independently of the others. "

The X-ray diffraction pattern of a pure substance is, therefore, like a fingerprint of the substance. The powder diffraction method is thus ideally suited for characterization and identification of polycrystalline phases. It is a versatile, non-destructive technique that reveals detailed information about the chemical composition and crystallographic structure of natural and manufactured materials.

Today about 50,000 inorganic and 25,000 organic single components, crystalline phases, diffraction patterns have been collected and stored on different databases. The main use of powder XRD is to identify components in a sample by a search/match procedure. Furthermore, the areas under the peak are related to the amount of each phase present in the sample.

Many solid state reactions produce polycristalline powders, therefore, for a high percentage of perovskites, the powder XRD is one of the main tools to characterize the material. Thus, it is indispensable to have a diffractometer in the laboratory.

Typically, a diffractometer consists of a tube of X-ray (source of radiation), a monochromator to choose the wavelength (for example, in the case of an anode of copper, the $K\alpha_2$ component could be eliminated by using a primary monochromator), slits to adjust the shape of the beam, a sample and a detector. In a more complicated apparatus also a goniometer can be used for fine adjustment of the sample and the detector positions. When an area detector is used to monitor the diffracted radiation a beamstop is usually needed to stop the intense primary beam that has not been diffracted by the sample. Otherwise the detector might be damaged. Usually the beamstop can be completely impenetrable to the X-ray or it may be semitransparent. The use of semitransparent beamstop allows the possibility to determine how much the sample absorbs the radiation using the intensity observed through the beamstop. Figure 9a) shows an X-ray powder diffractometer commonly seen in laboratories. Figure 9b) shows a detail of its geometry.

Ideally, every possible crystalline orientation is represented very equally in a powdered sample. The resulting orientational averaging causes the three-dimensional reciprocal space that is studied in single crystal diffraction to be projected onto a single dimension.

Bragg's law gives the angles for coherent and incoherent scattering from a crystal lattice. When X-rays are incident on an atom, they make the electronic cloud move as does any electromagnetic wave. The movement of these charges re-radiates waves with the same frequency (blurred slightly due to a variety of effects); this phenomenon is known as Rayleigh scattering (or elastic scattering). The scattered waves can themselves be scattered but this secondary scattering is assumed to be negligible. A similar process occurs upon scattering neutron waves from the nuclei or by a coherent spin interaction with an unpaired electron. These re-emitted wave fields interfere with each other either constructively or destructively (overlapping waves either add together to produce stronger peaks or subtract from each other to some degree), producing a diffraction pattern on a detector or film. The resulting wave interference pattern is the basis of diffraction analysis. This analysis is called Bragg diffraction.

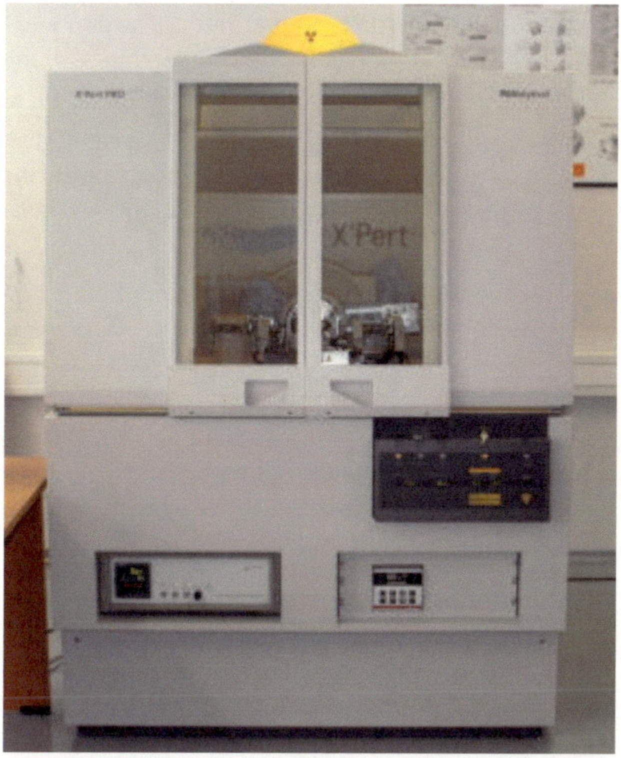

a)

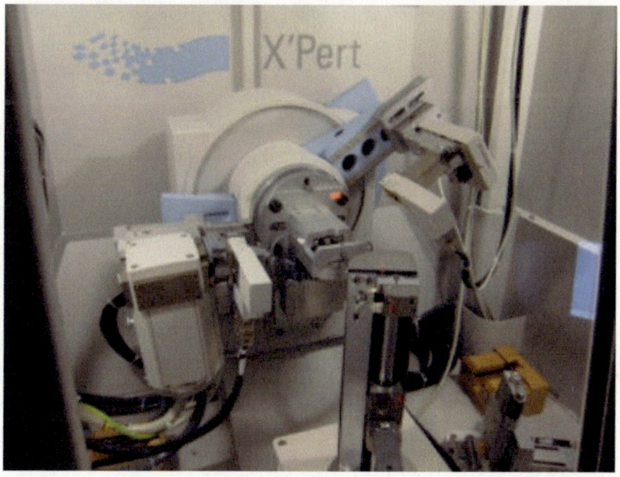

b)

Fig. 9. X-ray powder diffractometer.

Bragg diffraction (also referred to as the Bragg formulation of X-ray diffraction) was first proposed by William Lawrence Bragg and William Henry Bragg in 1913 in response to their discovery that crystalline solids produced surprising patterns of reflected X-rays (in contrast to that of, say, a liquid). They found that these crystals, at certain specific wavelengths and incident angles, produced intense peaks of reflected radiation (known as Bragg peaks). The concept of Bragg diffraction applies equally to neutron diffraction and electron diffraction processes. Both neutron and X-ray wavelengths are comparable with inter-atomic distances (~150 pm) and thus are an excellent probe for this length scale.

W. L. Bragg explained this result by modeling the crystal as a set of discrete parallel planes separated by a constant parameter *d*. It was proposed that the incident X-ray radiation would produce a Bragg peak if their reflections off the various planes interfered constructively.

The interference is constructive when the phase shift is a multiple of 2π; this condition can be expressed by Bragg's law (Equation 1),

$$n\lambda = 2d\,\mathrm{sen}\theta \tag{1}$$

where n is an integer, λ is the wavelength of the incident wave, d is the spacing between the planes in the atomic lattice, and θ is the angle between the incident ray and the scattering planes (Figure 10).

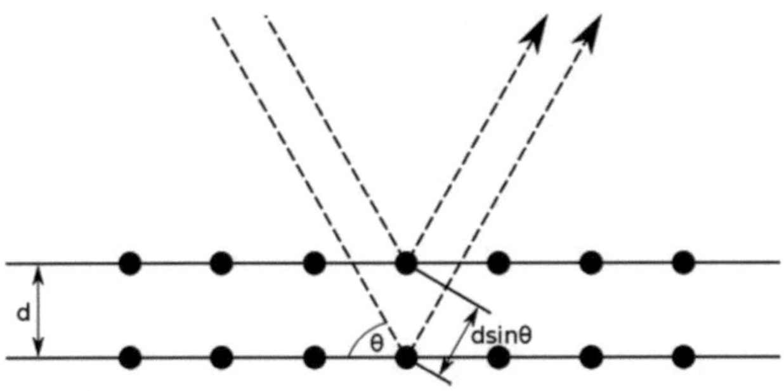

Fig. 10. Bragg diffraction.

The Rietveld method allows us to characterize the polycrystalline materials by a least squares approach to refine a theoretical line profile until it matches the measured profile shown in the pattern. The introduction of this technique was a significant step forward in the diffraction analysis of powder samples as, unlike other techniques at that time, it was able to deal reliably with strongly overlapping reflections.

The Rietveld refinement considers the fitting of a series of parameters linked, most of them, to the structure of the perovskite (or material in general). Some of them are the unit cell parameters, and the profile ones. It is important to reach a certain grade of reliability during the fitting process, and this can be followed by checking the so-called agreement factors (or reliability factors). Table 2 shows an example of lattice parameters and agreement factors obtained after the Rietveld refinement of the cubic perovskite-type solid solution $Ba_2In_{2-x}Co_xO_5$ ($0.50 \leq x \leq 1.70$). In Figure 11 an example of a Rietveld refined pattern, exactly for $Ba_2In_{0.3}Co_{1.7}O_5$, can be observed.

Compound	a (Å)	χ^2	R_p	R_{wp}	R_{exp}
$Ba_2In_{1.50}Co_{0.50}O_5$	4.2277(2)	5.02	13.2	18.1	8.07
$Ba_2In_{1.30}Co_{0.70}O_5$	4.1751(1)	3.58	16.5	21.0	11.1
Ba_2InCoO_5	4.1623(2)	3.67	12.4	15.3	8.01
$Ba_2In_{0.30}Co_{1.70}O_5$	4.1191(2)	2.28	9.93	13.1	8.65

Table 2. Cell parameters and reliability factors obtained from the Rietveld refinement.

In order to carry out a Rietveld refinement to fit the corresponding theoretical profile - model - a statistically acceptable pattern is necessary. The goodness of the pattern depends on the sample and the measurement parameters. There are several programs suitable to do Rietveld fittings, e.g. Fulprof and GSAS, among others.

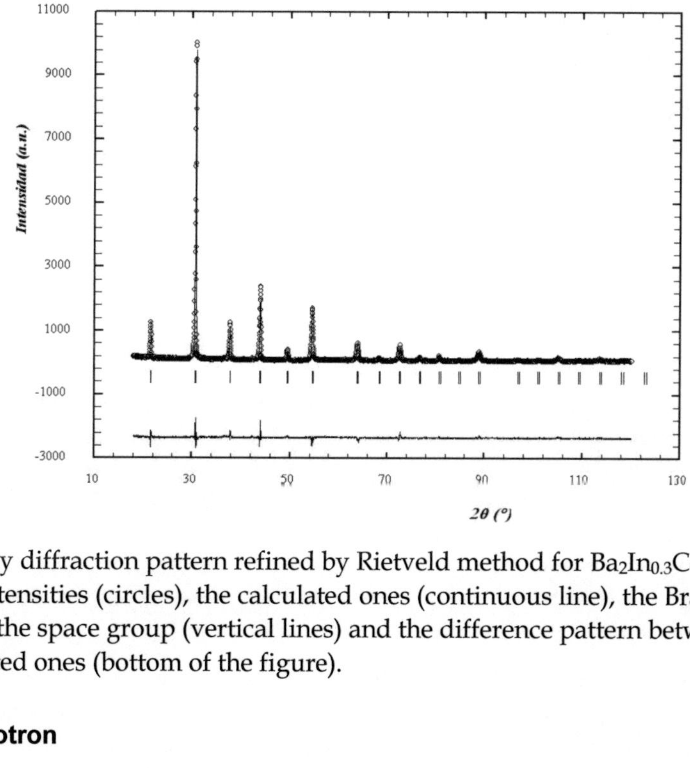

Fig. 11. X-ray diffraction pattern refined by Rietveld method for $Ba_2In_{0.3}Co_{1.7}O_5$, showing the observed intensities (circles), the calculated ones (continuous line), the Bragg positions allowed by the space group (vertical lines) and the difference pattern between the observed and calculated ones (bottom of the figure).

5.1 Synchrotron

A synchrotron is a particular type of cyclic particle accelerator in which the magnetic field (to turn the particles so they circulate) and the electric field (to accelerate the particles) arecarefully synchronized with the travelling particle beam. In essence, it is a X-ray source with variable wavelength. Thus, it allows going deeper in the X-ray diffraction analysis of the perovskites or any other material. There is a good number of synchrotron facilities all over the world. ESRF in France and Diamond Light Source in the UK constitute two examples (Figure 12).

Fig. 12. Diamond Light Source (UK).

6. Neutron diffraction

Neutron diffraction or elastic neutron scattering is the application of neutron scattering to the determination of the atomic and/or magnetic structure of a material. The sample is placed in a beam of thermal or cold neutrons to obtain a diffraction pattern that provides information of the structure of the material. The technique is similar to XRD but due to the different scattering properties of neutrons versus X-rays complementary information can be obtained.

A neutron diffraction measurement requires a neutron source (e.g. a nuclear reactor or spallation source), a sample (the perovskite to be studied or in general any material), and a detector. Sample sizes are large compared to those used in XRD. The technique is therefore mostly performed as powder diffraction. At a research reactor other components such as crystal monochromators or filters may be needed to select the desired neutron wavelength. Some parts of the setup may also be movable. At a spallation source the time of flight technique is used to sort the energies of the incident neutrons (higher energy neutrons are faster), so no monochromator is needed, but rather a series of aperture elements synchronized to filter neutron pulses with the desired wavelength.

Neutron diffraction is closely related to XRD. In fact the single crystal version of the technique is less commonly used because currently available neutron sources require relatively large samples and large single crystals are hard or impossible to come by for most materials. Future developments, however, may well change this picture. Because the data is typically a 1D powder pattern they are usually processed using Rietveld refinement. In fact the latter found its origin in neutron diffraction (at Petten in the Netherlands) and was later extended for use in XRD.

One practical application of elastic neutron scattering/diffraction is that the lattice constant of perovskites and other crystalline materials can be very accurately measured. Together with an accurately aligned micropositioner a map of the lattice parameters through the material can be derived. This can easily be converted to the stress field experienced by the compound. This has been used to analyze stresses in aerospace and automotive components to give just two examples. There is a good number of facilities all over the world offering a neutron source. Among them, we could mention, for instance, ISIS in the UK, and ILL in France.

Figure 13 shows a picture of the reactor hall at ILL in Grenoble, France. ILL (Institut Laue-Langevin) is one the most important centres in the world to carry out neutron experiments.

Fig. 13. Inside the reactor hall at ILL in Grenoble (France).

7. High-resolution transmission electron diffraction

The original form of electron microscope, the transmission electron microscope (TEM) uses a high voltage electron beam to create an image. The electron beam is accelerated by an anode with respect to the cathode, focused by electrostatic and electromagnetic lenses, and transmitted through the sample that is in part transparent to the electrons and in part scatters them out of the beam. When it emerges from the sample, the electron beam carries information about the structure of the sample that is magnified by the objective lens system of the microscope. Hardware correction of spherical aberration for the high-resolution transmission electron microscopy (HRTEM) has allowed the production of images with resolution below 0.5 Å at magnifications above 50 million times. This possibility of having direct images of the atomic arrangement in the structure has made the HRTEM an important tool for nano-technologies research and development.

8. Conclusion

The perovskite is a mineral series composed of calcium titanate. Many transition metal oxides show that very versatile perovskite structure. The rich variety of physical properties such as high-temperature superconductivity and colossal magnetoresistance observed in these compounds makes them very attractive from both fundamental and applied perspectives.

There are different methods of synthesis. Among them, one of the most used is that known as ceramic method, although not the most efficient one. The freeze-drying method offers purer materials by reducing the heating time and working at not so high temperatures.

The general chemical formula for perovskite compounds is ABX_3, where *A* and *B* are two cations of very different sizes, and X is an anion that bonds to both. The *A* atoms are larger than the *B* atoms. The ideal cubic-symmetry structure has the *B* atoms in 6-fold coordination, surrounded by an octahedron of anions, and the *A* atoms in 12-fold cuboctahedral coordination. The relative ion size requirements for stability of the cubic structure are quite stringent, so slight buckling and distortion can produce several lower-symmetry distorted versions, in which the coordination numbers of *A* cations, *B* cations or both are reduced. The orthorhombic and tetragonal phases are the most common non-cubic variants.

The X-ray diffraction pattern of a pure substance is like a fingerprint of the substance. The powder diffraction method is thus ideally suited for characterization and identification of polycrystalline phases, and therefore of perovskites.

The Rietveld method allows us to characterize the polycrystalline materials by a least squares approach to refine a theoretical line profile until it matches the measured profile shown in the pattern.

It is possible to go further in the study of the structure of the perovskites by means of neutron diffraction, although it is true to say that large facilities are needed to carry out such study.

9. Acknowledgment

The author of this chapter thanks Professors Pedro Núñez and Cristina González-Silgo at Universidad de La Laguna (Spain), and Professor John E. Greedan at McMaster University (Canada) for such important contributions to the field of the perovskites which came strongly in useful for the development of his knowledge of it as well as for having been remarkable mentors during his years of research.

10. References

Bragg, W.L. (1913). The diffraction of short electromagnetic waves by a crystal, *Proceedings of the Cambridge Philosophical Society*, Vol.17, pp. 43-57

Coey, J.M.D., Viret, M., von Molnar, S. (1999). Mixed-valence manganites, *Advances in Physics*, Vol.48 (2), pp. 167–293. doi:10.1080/000187399243455

Cullity, B.D. (1977). Elements of X-ray diffraction, Addison-Wesley, 2nd ed., ISBN 0-201-01174-3

Didenko, Y.T., Suslick, K.S. (2002). The energy efficiency of formation of photons, radicals and ions during single-bubble cavitation. *Nature*, Vol.418, pp. 394-397

Ekambaram, S., Patil, K.C., Maaza, M. (2005). Synthesis of lamp phosphors: facile combustion approach. *J. Alloys Comp.*, Vol.393, pp. 81–92

Gedanken, A. (2004). Using sonochemistry for the fabrication of nanomaterials. *Ultrasonics Sonochemistry*, Vol.11, pp. 47-55

Gómez-Cuaspud, J.A., Valencia-Ríos, J.S., Carda-Castellón, J.B. (2010). Preparation and characterization of perovskite oxides by polymerization-combustion, *J. Chil. Chem. Soc.*, Vol.4, pp. 55

Goodenough, J.B. (2004). *Rep. Prog. Phys.*, Vol.67, pp. 1915

Ibberson, R.M., David, W.I.F. (2002). Structure determination from neutron powder diffraction data, *IUCr monographs on crystallography*, Oxford scientific publications, ISBN 0-19-850091-2

Larson, A.C., Von Dreele, R.B. (2000). General Structure Analysis System (GSAS). *Los Alamos National Laboratory Report* LAUR 86-748

Lozano-Gorrín, A.D., Greedan, J.E., Núñez, P., González-Silgo, C., Botton, G.A., Radtke, G. (2007). Structural characterization, magnetic behavior and high-resolution EELS study of new perovskites $Sr_2Ru_{2-x}Co_xO_6$ ($0.5 \leq x \leq 1.5$), *Journal of Solid State Chemistry*, Vol.180, pp. 1209-1217

Lozano-Gorrín, A.D., Núñez, P., López de la Torre, M.A., Romero de Paz, J., Sáez-Puche, R. (2002). Spin glass behavior of new perovskites $Ba_2In_{2-x}Co_xO_5$ ($0.5 \leq x \leq 1.70$), *Journal of Solid State Chemistry*, Vol.165, pp. 254-260, doi:10.1006/jssc.2002.9515, available online at http://www/idealibrary.com

Megaw, H.D. (1945). Crystal structure of barium titanate, *Nature*, Vol.155, pp. 484-485, ISSN 00280836

Navrotsky, A. (1998). Energetics and crystal chemical systematics among ilmenite, lithium niobate, and perovskite structures, *Chem. Mater.*, Vol.10, pp. 2787. doi:10.1021/cm9801901

Patil, K.C., Aruna, S.T., Mimani, T. (2002). Combustion synthesis: an update. *Solid State Mater. Sci.*, Vol.6, pp. 507–512

Pechini, M.P. (1967). US Patent No.3.330.697 July 1

Rao, C.N.R., Gopalakrishnan, J. (1997). New directions in Solid State Chemistry. Cambridge U. Press, ISBN 0-521-49559-8

Reisfeld, R., Jorgensen, C.K. (1992). Optical properties of colorants or luminescent species in sol-gel glasses, structure and bonding. Springer-Verlag, Heidelberg

Rodríguez-Carvajal, J. (1993). Recent advances in magnetic structure determination by neutron powder diffraction, *Physica B*, Vol.192, pp. 55

Roisnel, T., Rodríguez-Carvajal, J. (2000). WinPLOTR: a Windows tool for powder diffraction patterns analysis. *Proceedings of the Seventh European Powder Diffraction Conference (EPDIC 7)*, pp. 118-123, Ed. R. Delhez and E.J. Mittenmeijer

Rose, H.H. (2008). Optics of high-performance electron microscopes. *Sci. Technol. Adv. Mater.*, Vol.9, pp. 014107

Toby, B.H. (2001). EXPGUI, a graphical user interface for GSAS, *J. Appl. Cryst.*, Vol.34, pp. 210-213

Witze, A. (2010). Building a cheaper catalyst. *Science News* Web Edition. http://www.sciencenews.org/view/generic/id/57618/title/Building_a_cheaper_catalyst

6

Controlled Crystallization of Isotactic Polypropylene Based on α/β Compounded Nucleating Agents: From Theory to Practice

Zhong Xin and Yaoqi Shi
State Key laboratory of Chemical Engineering, College of Chemical Engineering, East China University of Science and Technology, Shanghai China

1. Introduction

Isotactic polypropylene (iPP) is one of the most important thermoplastic polymers owing to its low manufacturing cost and versatile properties. Moreover, iPP exhibits a very interesting polymorphic behavior (Awaya, 1988; Busse et al., 2000; Lotz et al., 1996; Vagar, 1992). At least five modifications: monoclinic α form, trigonal β form, orthorhombic γ form, δ and smectic phase have been reported. The α form is the best known and most stable in commercial grades of iPP being found in most melt crystallized specimens, especially those being added α Nucleating agents (NA) (Labour et al., 1999; Vagar, 1986). The β form is metastable thermodynamically and is obtained under some special conditions such as a high degree of supercooling, temperature gradient, shear-induced crystallization or addition of β-nucleating agents (Fillon et al., 1993; Ismail & Al-Raheil, 1998). The γ form occurs in low-molecular-weight iPP or under high pressure and the mesomorphic form results from quenching (Meille et al., 1990; Lotz et al., 1986). Different crystalline form of iPP leads different properties like optical and mechanical properties.

NA as one of the additives presents a role of increasing the nucleation density of polymer greatly and enhancing the nucleation rate dramatically so as to have a great impact on the mechanical properties of polymer (Kristiansen et al., 2003; Romankiewicz et al., 2004; Tenma & Yamaguchi, 2007). So far, two kinds of NAs, α phase and β phase NAs discriminated by the form of iPP they induce have been widely put into use in modifying iPP. The α phase NA can improve the stiffness and optical properties of iPP while decrease its toughness (Gui et al., 2003; Zhang G.P. et al., 2003; Zhang Y.F. & Xin, 2006). The β phase NA will induce β-iPP during crystallization, which can improve toughness and heat distortion temperature of iPP while decrease its stiffness (Tordjeman et al., 2001; Zhao et al., 2008). Thereby, it is well expected to balance the iPP's stiffness and toughness. Xin's research group firstly proposed the idea of compounding α/β NAs. However, whether compounding α, β NAs will enhance stiffness and toughness simultaneously or not and what influence will α/β compounded NAs take on the crystallization kinetics, crystallization morphologies, and mechanical proprieties of iPP call our eye.

In this work, three kinds of well studied α/β compounded NAs, Phosphate/Amide, Sorbitol/Amide, and Phosphate/Carboxylate were selected to review. This short review aims to present some conclusions of α/β compounded NAs and to lay the foundation for compounding α and β NAs afterwards.

2. Crystallization kinetics of iPP nucleated with α/β compounded NAs

Crystallization process of semi crystalline polymers such as polypropylene can have a dramatic impact on the mechanical properties. Thus, we studied the crystallization kinetics of iPP nucleated by α/β compounded NAs first.

Isothermal crystallization kinetics of iPP nucleated with Phosphate/Amide compounded NA, NA40/NABW was studied by Zhang et al. (Zhang & Xin, 2007). The results showed that Avrami equation, as shown below, was quite successful for analyzing the experimental data of the isothermal crystallization kinetics.

$$1 - X_t = \exp(-Z_t t^n) \tag{1}$$

where X_t is the relative crystallinity at time t, n is Avrami exponent, a constant whose value depends on the mechanism of nucleation and on the form of crystal growth, and Z_t is a constant containing the nucleation and growth parameters. The addition of NA40/NABW could shorten crystallization half-time ($t_{1/2}$) and increase crystallization rate of iPP greatly. Consequently the molding cycle time of iPP would be reduced obviously, which has great importance for polymer processing. The Avrami exponents of iPP and nucleated iPP were close to 3, indicating that the addition of nucleating agents did not change the crystallization growth patterns of iPP under isothermal conditions and the crystal growth was heterogeneous three-dimensional spherulitic growth. The Caze method was applied to study on the non-isothermal crystallization kinetics of nucleated iPP by Phosphate/Amide compounded NA, NA11/DCHT (Zhao & Xin, 2010). It can be seen from the results that the addition of the α/β compounded NAs can obviously shorten $t_{1/2}$ of iPP, especially at lower cooling rates. When the cooling rate Φ is 2.5°C/min, $t_{1/2}$ of nucleated iPP was 104.9s, while that of pure iPP was 135.4 s. The Avrami exponent n for nucleated iPP indicated that the α/β compounded NA acted as heterogeneous nuclei followed by three-dimensional spherical growth during non-isothermal crystallization. Therefore, the type of nucleation of iPP was significantly changed in the presence of the α/β compounded NAs while the geometry of crystal growth of iPP did not change.

Bai et al. investigated the isothermal crystallization kinetics of nucleated by Sorbitol/ Amide compounded NA, DMDBS/TMB-5 (Bai & Wang, 2009). The crystallization kinetics parameters suggested that compounded NA accelerated the crystallization process of iPP greatly. $t_{1/2}$ of iPP/DMDBS/TMB-5 was much smaller than iPP, indicating the faster crystallization process by the addition of compounded NA. For all the samples, the Avrami exponent value n ranges from 2 to 3, which means spherulite development arose from an athermal heterogeneous nucleation. The fold surface free energy of virgin iPP and nucleated iPP was also calculated from the crystallization kinetics. Samples with addition of compounded NA resulted in smaller values. That means interfacial surface free energy of iPP was reduced with the presence of compounded NA. Similar results were obtained by the study on the non-isothermal crystallization kinetics of iPP nucleated by Sorbitol/Amide compounded NA, 3988/DCHT (Zhao & Xin, 2010).

Except for Amide NA, Carboxylate NA is proved to be another highly effective β NA for iPP. Xu gave us the picture of non-isothermal crystallization kinetics of iPP nucleated by Phosphate/Carboxylate compounded NA, NA40/H-Ba (Xu, 2010). From the point view of crystallization temperature, the addition of NA40/H-Ba enhanced the crystallization rate of iPP. Judging from the Avrami exponent, the spherulite of iPP grew in the way of three-dimensional during non-isothermal crystallization with the presence of NA40/H-Ba, which was in accordance with the other two α/β compounded NAs.

	Compounded NAs	T_C/°C	$t_{1/2}$/s	n	$\bar{n}$
		123	42	2.70	
		125	66	2.58	
	iPP	127	108	2.84	2.85
		129	197	3.10	
		131	428	3.04	
Phosphate/Amide		133	24	2.76	
	NA40/NABW	135	47	2.61	2.77
		137	78	2.77	
		149	141	2.99	
		141	252	2.72	
		124	180	2.44	
		126	306	2.43	
	iPP	128	516	2.6	2.54
		130	840	2.63	
		132	1356	2.58	
Sorbitol/Amide		134	30	2.65	
	DMDBS/TMB-5	136	54	2.90	2.96
		138	102	3.00	
		140	204	3.30	

Table 1. Isothermal crystallization kinetics parameters of pure iPP and nucleated iPP (Bai & Wang, 2009; Zhang & Xin, 2007)

	Compounded NAs	Cooling rate Φ/(°C/min)	T_C/°C	$t_{1/2}$/s	n
Phosphate/Amide	iPP	2.5	121.4	135	3.75±0.03
		5	118.8	78	
		10	115.9	44	
		20	112.7	23	
		40	108.9	14	
	NA11/DCHT	2.5	133.7	104	3.66±0.11
		5	131.2	60	
		10	128.5	25	
		20	125.6	16	
		40	121.7	8	
Sorbitol/Amide	iPP	2.5	121.4	135	3.75±0.03
		5	118.8	78	
		10	115.9	44	
		20	112.7	23	
		40	108.9	14	
	3988/DCHT	2.5	128.7	121	2.88±0.25
		5	124.6	69	
		10	120.2	43	
		20	115.4	27	
		40	111.9	17	
Phosphate/Carboxylate	iPP	2.5	127.8	127	3.67±0.09
		5	124.7	124	
		10	121.6	121	
		15	119.8	119	
		20	118.4	118	
	NA40/H-Ba	2.5	137.9	115	4.52±0.04
		5	135.3	60	
		10	132.6	33	
		15	130.8	23	
		20	129.7	18	

Table 2. Non-isothermal crystallization kinetics parameters of pure iPP and nucleated iPP (Xu, 2010; Zhao & Xin, 2010)

Isothermal and non-isothermal crystallization kinetics of iPP nucleated by three kinds of α/β compounded NAs were reviewed in this section. It can be concluded that compounded

NAs will increase the crystallization temperature of iPP, shorten the crystallization half-time. Consequently the molding cycle time of iPP will be reduced obviously, which has great importance for polymer processing. The obtained Avrami exponents indicated that the type of nucleation of iPP is changed from homogeneous to heterogeneous in the presence of the α/β compounded NAs while the geometry of crystal growth of iPP remains three-dimension spherical growth.

3. Crystallization morphologies of iPP nucleated with α/β compounded NAs

The spherulite size of iPP can be decreased by cooperation with any kinds of NAs. But the morphology of nucleated iPP largely depends on the types of NA. The α NA will only induce α form iPP while β form iPP can be obtained by incorporating with β NA. Then what about the morphologies of iPP nucleated by α/β compounded NAs?

Polarized optical microscope was used to investigate the crystallization morphologies of iPP nucleated with Phosphate/Carboxylate compounded NA, NA40/H-Ba by Xu et al. (Xu et al., 2011). As shown in Fig.1, in nucleated iPP, a large number of nuclei would be produced due to the existence of NAs. Therefore the spherulites cannot grow large enough to overlap, the size of spherulites in nucleated iPP would be much smaller than those in pure iPP. But as to the morphologies of the samples, iPP nucleated with NA40/H-Ba showed no sign of bright and colorful β crystals, appeared much close to the morphology of iPP induced by NA40 individually.

Fig. 1. Polarized light microphotographs for pure iPP and nucleated iPP samples crystallized at 135 °C (Xu et al., 2011)

The crystallization morphologies of pure iPP and iPP induced by Sorbitol/Amide compounded NA, 3988/DCHT were shown in Fig.2 (Zhao & Xin, 2010). From figure, it can be seen that with the addition of the α/β compounded NA, the spherulite size decreased significantly. Different from iPP with NA40/H-Ba, β form iPP became the majority in the morphology of iPP nucleated with 3988/DCHT. It can be considered that DCHT played a leading role during crystallization. The same conclusion was drew by investigating of iPP cooperation with 3988/NABW (Xu, 2010). The morphology of nucleated iPP was close to that incorporation with NABW individually. In addition, Bai et al. directly observed the crystallization morphologies of iPP nucleated with Sorbitol/Amide compounded NA, DMDBS/TMB-5 by SEM (Bai et al., 2008). Pure iPP showed the growth of well developed α spherulites with 30~50um in diameter. The size of iPP spherulites was also reduced with addition compounded NA. Similarly β form iPP dominated in the morphology of PP/0.1DM/0.1TM.

Fig. 2. Polarized light microphotographs for pure iPP and nucleated iPP samples crystallized at 140 °C (Zhao & Xin, 2010) (a) pure iPP, (b) iPP/ (3988/DCHT)

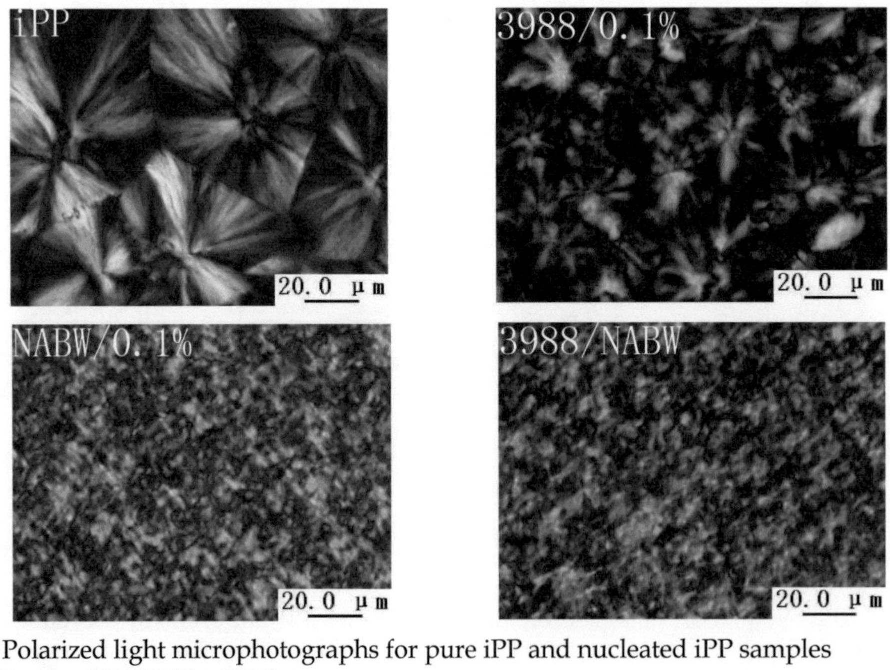

Fig. 3. Polarized light microphotographs for pure iPP and nucleated iPP samples crystallized at 135 °C (Xu, 2010)

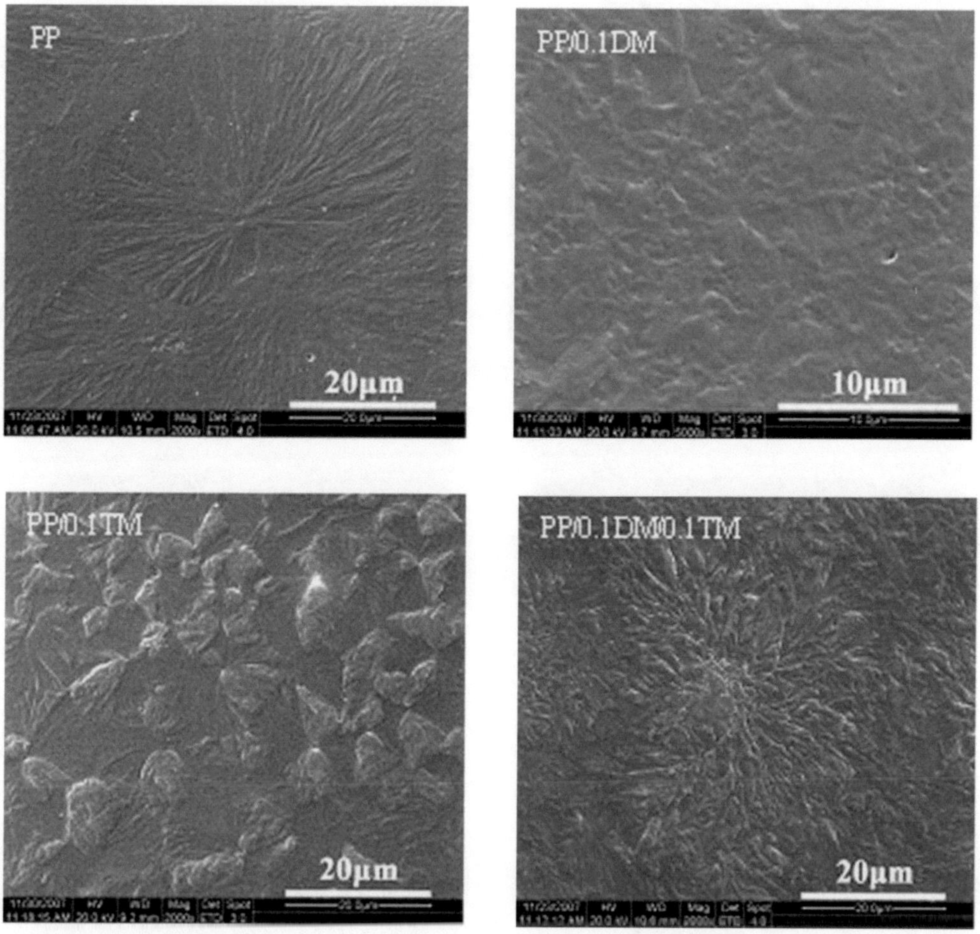

Fig. 4. SEM for pure iPP and nucleated iPP samples (Bai et al., 2008)

However, it was interesting that totally different results could be gained when the DCHT compounded with different α NAs. Zhao et al. found the nucleation effect of NA11/DCHT compounded NA was between that of iPP/NA11 and iPP/DCHT. It can be seen from Fig.5, the spherulites of pure iPP showed the typical characteristic of α crystal, which had a large size and clear boundaries (Zhao & Xin, 2010). By adding compounded NA, the spherulite size greatly reduced, indicating that compounded NA played a role of heterogeneous nuclei during crystallization. The content of bright and colorful β form iPP was less than that of iPP/DCHT, but was higher than iPP/NA40, which means at this condition the crystallization morphology of iPP was affected by both NAs within the compounded NA. The same result was got through the study on crystallization morphologies of iPP nucleated with Phosphate/Amide compounded NA, NA40/NABW by Xu, as shown in Fig.6 (Xu, 2010).

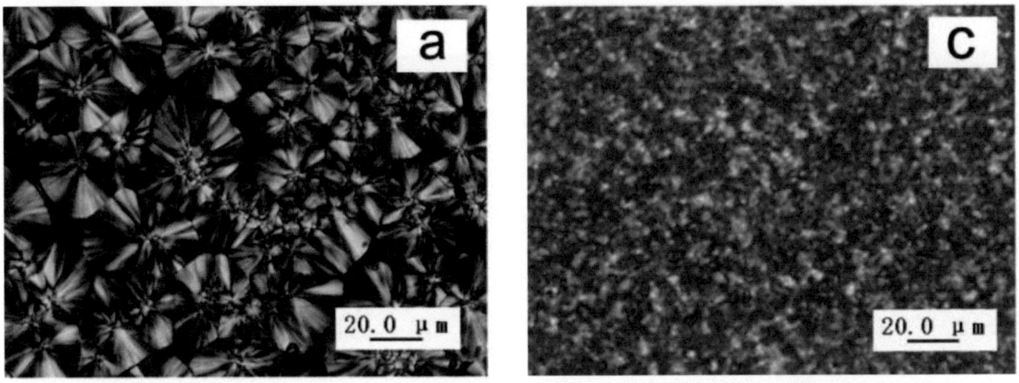

Fig. 5. Polarized light microphotographs for pure iPP and nucleated iPP samples crystallized at 140 °C (Zhao & Xin, 2010) (a) iPP, (c) iPP/ (NA11/DCHT)

Fig. 6. Polarized light microphotographs for pure iPP and nucleated iPP samples crystallized at 135 °C (Xu, 2010)

All studies showed that the size of spherulites in nucleated iPP appeared much smaller than that of in pure iPP. However, iPP nucleated by different α/β compounded NAs showed different crystallization morphologies. The morphology of iPP nucleated Phosphate/Carboxylate compounded NA, NA40/H-Ba was close to iPP nucleated by NA40 individually, while iPP nucleated by Sorbitol/Amide compounded NA, 3988/DCHT showed the similar morphology of iPP/DCHT. IPP nucleated with Phosphate/Amide NA40/NABW compounded NA presented a crystallization morphology that combined both NAs' within the compound system.

4. Mechanical properties of iPP nucleated with α/β compounded NAs

The effects of different NAs on the crystallization process of polymer reflect on its mechanical properties finally, which determines on the use value of the polymer. Mentioned in the introduction, the α NA can improve the stiffness and optical properties of iPP while decrease its toughness. The β NA will induce β-iPP during crystallization, which can improve toughness and heat distortion temperature of iPP while decrease its stiffness. Will it come true that we can balance the iPP's stiffness and toughness by compounding two kinds of nucleating agents?

Xu et al. investigate the effect of Phosphate/Carboxylate α/β compounded NA, NA40/HB on the mechanical properties of iPP (Xu et al., 2011). As shown in Tab.3, tensile strength (ASTM D-638) and flexural modulus (ASTM D-790) of iPP were improved with the presence of NA40 while the impact strength (ASTM D-256) decreased. On the contrary, the impact strength of iPP could increase to 3.4 times to that of pure iPP but tensile strength and flexural modulus was reduced as always by adding HB. Numerically the mechanical properties of iPP nucleated with NA40/HB were close to that iPP/NA40, which showed no sign of enhancing the toughness of iPP.

	Compounded NA	Tensile strength /MPa	Flexural modulus /MPa	Impact strength /(J/m)
Phosphate/Amide	iPP	29.8	1223	33.8
	NA11 (0.1 wt %)	34.5	1770	30.2
	DCHT (0.1 wt %)	27.9	1143	74.0
	NA11/DCHT (1:1)	34.2	1669	49.7
Sorbitol/Amide	iPP	29.8	1223	33.8
	3988 (0.1 wt %)	31.5	1297	30.9
	DCHT (0.1 wt %)	27.9	1143	74.0
	3988/DCHT (1:1)	27.6	1108	73.4
Phosphate/Carboxylate	iPP	33.1	1052	35.6
	NA40 (0.1 wt %)	36.2	1562	33.2
	HB (0.1 wt %)	28.8	1025	158.2
	NA40/HB (1:1)	36.3	1521	34.2

Table 3. Mechanical Properties of Pure iPP and iPP Nucleated with Individual α, β and α/β Compounded NAs (Xu et al., 2011; Zhao & Xin, 2010)

Similar to the effect on crystallization morphologies, different mechanical properties of iPP would be reached when the DCHT compounded with different α NAs. Incorporation with Sorbitol/Amide compounded NA, 3988/DCHT can significantly improve the impact strength of iPP, but was not benefit to the stiffness like tensile strength, flexural modulus (Zhao & Xin, 2010). It is exciting that the goal of enhancing the stiffness and toughness of iPP simultaneously can be reached by compounding DCHT with another α NA, NA11. The tensile strength, flexural modulus and impact strength of iPP nucleated with NA11/DCHT was higher than those of pure iPP.

Mechanical properties such as tensile strength, flexural modulus and impact strength of iPP nucleated with three kinds of α/β compounded NAs were investigated. The similar results to the study on crystallization morphologies were obtained. Incorporation with Phosphate/Carboxylate compounded NA, NA40/H-Ba only enhanced the stiffness of iPP, while with Sorbitol/Amide compounded NA, 3988/DCHT increased the toughness of iPP, which was close to iPP nucleated by DCHT individually. Compounding NA40 and DCHT could reach a good balance between stiffness and toughness of iPP. Then, what factor plays a dominant role when compounding α, β two kinds of nucleating agents?

5. Optimization method for compounding α, β NAs

Through reviewing on the crystallization kinetics, crystallization morphologies, and mechanical proprieties of iPP nucleated by α/β compounded NAs, it can be noticed that some α, β NAs can induce iPP during crystallization respectively when they are compounded, hence improve the stiffness and toughness simultaneously. While some α or β NA will play a leading role when it compounds with another NA. Thus the nucleating effect of the compounded NA appears close to the leading one, which goes against original intention of compounding α and β NA. So find out the key factor of affecting the effect of α/β compounded NAs is the precondition of successfully compounding α and β NA.

From the traditional crystallization point of view, the overall crystallization rate depends on two stages: nucleation and growth. In the nucleation process, the formation of nucleus relies on the molecular movement in the molten spontaneously. Once the nucleus came into existence, the crystal grows in the form what nucleus is. So nucleation is the precondition for crystallization. The role of NA is to provide a large number of nuclei before the self-nuclei formed, which results in changing the homogeneous nucleation into a heterogeneous one. Furthermore, several studies on crystallization kinetics show that the NA has little impact on the growth stage of crystallization (Cai et al., 2010; Huang et al., 2005; Zhao & Xin, 2010). Accordingly, we believe that the crystallization form of polypropylene depends on the NA which comes into effect first in the nucleation stage. This sequence can be judged by the crystallization temperature (Tc) of polypropylene nucleated with NA individually. The NA with higher Tc means earlier the NA nuclei could be "accepted" by polypropylene and consequently comes into effect first in compounded system. So Tc is considered as the key factor of affecting the effect of α/β compounded NAs.

Zhao et al. confirmed that the effect of compounded NA depends on which NA come into effect first during the nucleation stage (Zhao & Xin, 2010). The Tc of iPP induced by different NA individually was illustrated in Fig.7. It can be seen that Tc of iPP induced by DCHT was much higher than that of 3988. According to the mentioned assumption, when DCHT compounds with 3988, DCHT would play a leading role. Refers to the results in section 3 and section 4, it is clear that the nucleating effect such as crystallization morphologies and mechanical properties of Sorbitol/Amide compounded NA, 3988/DCHT appeared close to that of DCHT. It can be noticed from Fig.7 that Tc of NA11 showed little difference to that of DCHT. During crystallization competitive nucleation takes place between two NAs, which results in the combined crystallization morphology and simultaneously increasing tensile strength, flexural modulus and impact strength of iPP. Xu et al. came to the same result by investigating Phosphate/Carboxylate compounded NA, NA40/H-Ba (Xu et al., 2011). NA40, the one with higher Tc plays a leading role in the crystallization while H-Ba showed no effect on inducing iPP. Therefore Tc is proved to be the key factor of affecting the effect of α/β compounded NAs.

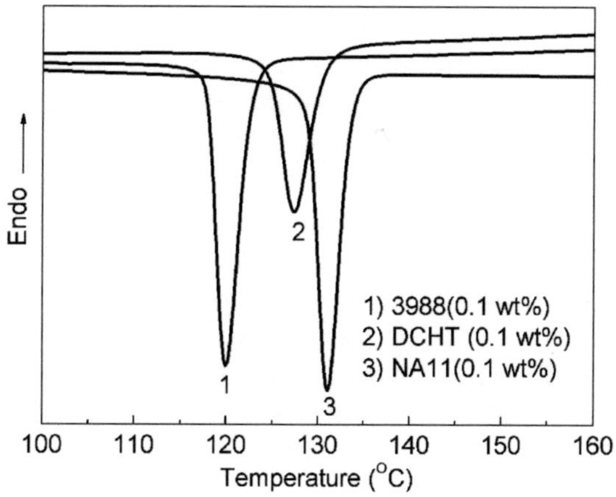

Fig. 7. DSC melting curves of iPP nucleated with individual α or β NAs (Zhao & Xin, 2010)

Sample	Tc / ℃
iPP	121.6
NA40/iPP	130.2
HB-a/iPP	125.1

Table 4. Crystallization temperature of iPP nucleated with different NAs (Xu et al., 2011)

The key factor of affecting the α/β compounded NAs was summarized in this part. That is the crystallization temperature of polypropylene nucleated with NA individually. The NA with higher Tc plays a leading role in the crystallization process. Consequently the mechanical properties, crystallization properties and crystallization morphologies of iPP appear close to it. Competitive nucleation will occur when the difference of Tc between the two NAs is not significant.

According to this, Tc becomes the one we can adjust that controls the crystallization behaviors of iPP based on α/β compounded NA. It can be easily conclude that the principle of compounding α and β NA is to make Tc of two NAs as close as possible, so as to let competitive nucleation happen. As known to all, the Tc of a NA depends on not only the species but the content of it as well. That is to say various addition amount of the same NA leads different Tc. Then, method for compounding α and β NAs can be developed according to this: First, we shall obtain Tc of iPP nucleated with different addition amounts of α and β NAs individually by DSC. Then list $T_{C\,\alpha}$ and $T_{C\,\beta}$ at each ratio of compounded α/β with a fixed concentration. The ratio which contains $T_{C\,\alpha} = T_{C\,\beta}$ will be the optimal compounded ratio of these two α and β NAs at this concentration. In this context, competitive nucleation will occur during crystallization.

6. Practice of adjusting the stiffness and toughness of iPP based on α/β compound NAs

Here an example of adjusting the stiffness and toughness of isotactic polypropylene based on different of α/β compound NAs was employed. Shi et al. studied the different ratios α/β compounded NAs on mechanical properties of iPP (Shi & Xin, 2011). It was found that stiffness and toughness of iPP could be adjusted and enhanced simultaneously by changing the ratio of α and β nucleating agents, as shown in Fig.8. Comparing to the others, the absolute value of difference of crystallization peak temperature between two kinds of NAs at optimal compounded ratio was the smallest. It verifies that the key factor summarized before can also be applied to different ratios α/β compounded NAs. Then relying on the established method, the optimal compounded ratios of NA40/H-Ba and NA40/PA-03 (PA-03, Carboxylate β NA for iPP) were obtained, at which there appeared $T_{C\,\alpha} = T_{C\,\beta}$ as shown in Fig.9. Refer to Fig.10, the calculated results were proved to be valid by the investigation of the effect of NA40/H-Ba and NA40/PA-03 with different ratio on mechanical properties of iPP, which means the method is applicable for compounding any NAs. Furthermore compounded NAs could enhance stiffness and toughness of iPP simultaneously with these ratios, as can be seen from Tab.5.

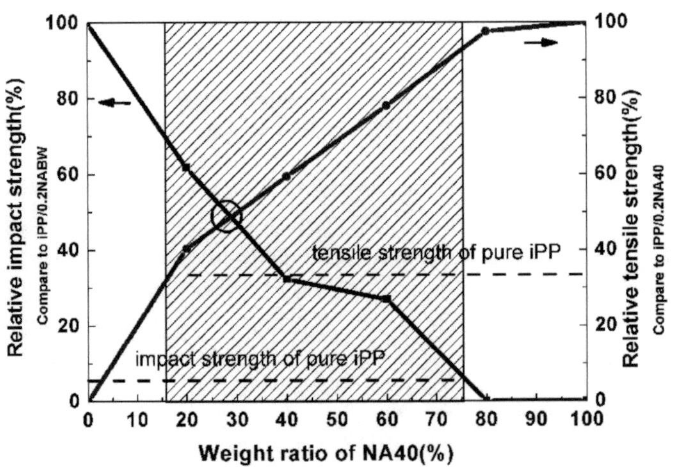

Fig. 8. Effect of NA40/NABW compounded NAs with different ratio on mechanical properties of iPP (addition amount 0.2wt %) (Shi & Xin, 2011)

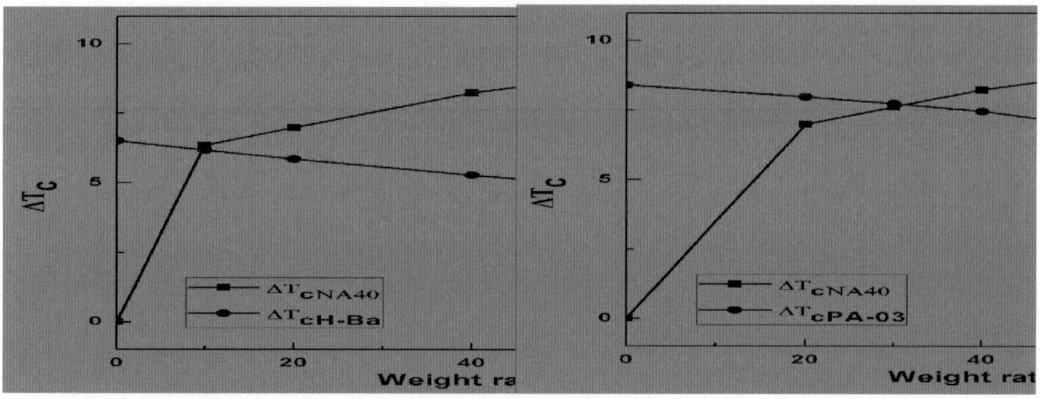

Fig. 9. $T_{C\alpha}$ and $T_{C\beta}$ at different ratio of NA40/H-Ba and NA40/PA-03 compounded NAs (addition amount 0.2 wt %)

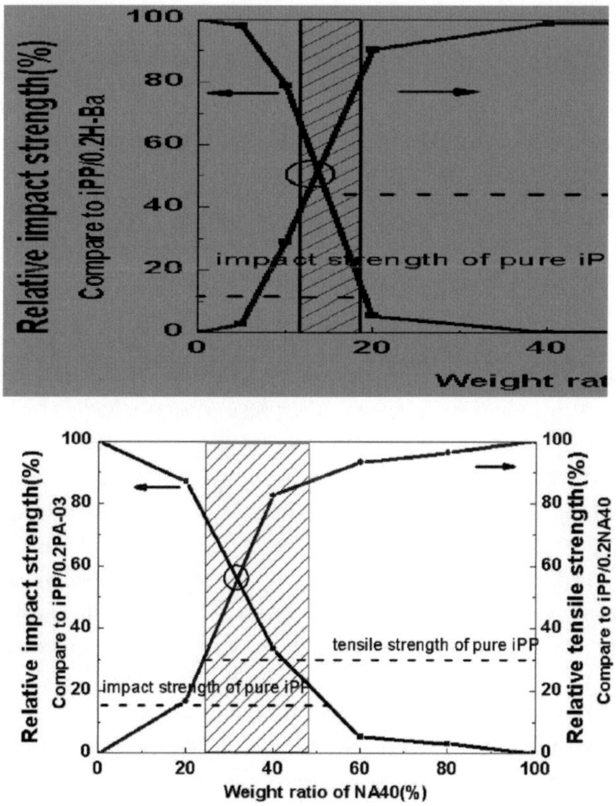

Fig. 10. Effect of NA40/H-BA and NA40/PA-03 compounded NAs with different ratio on mechanical properties of iPP (addition amount 0.2wt %) (Shi & Xin, 2011)

Nucleating agents	Compound ratio	Tensile strength (MPa)	Flexural modulus (MPa)	Impact strength (J/m)
iPP	-	29.9	1171	31.7
NA40	-	32.9	1616	27.1
NABW	-	25.4	1031	98.2
NA40/NABW	1:3	31.1	1443	58.0
H-Ba	-	26.9	1018	93.6
NA40/H-Ba	1:9	30.8	1309	79.0
PA-03	-	27.1	1022	103.7
NA40/PA-03	3:7	31.0	1344	64.0

Table 5. Mechanical Properties of iPP Nucleated with different NAs (addition amount 0.2 wt %) (Shi & Xin, 2011)

7. Conclusions

Nowadays α/β compounded NAs for polypropylene have attracted more and more attention. This short review summarized the research on α/β compounded NAs in recent years. Three kinds of well studied α/β compounded nucleating agents (NAs), Phosphate/Amide, Sorbitol/Amide, and Phosphate/Carboxylate were selected to review their influence on the crystallization kinetics, crystallization morphologies, and mechanical proprieties of isotactic polypropylene (iPP). The results showed that α/β compounded NAs could not only increase the crystallization temperature of iPP but also shorten the crystallization half-time, consequently reduce molding cycle time of iPP more obviously. The obtained Avrami exponent indicated that the type of nucleation of iPP could be changed while the geometry of crystal growth of iPP remains. The size of spherulites in nucleated iPP appeared much smaller than that in pure iPP. However, iPP nucleated by different α/β compounded NAs showed different morphologies. The same result was obtained by the investigation of the mechanical properties of iPP. Some α/β compounded NAs were able to enhance stiffness and toughness of iPP simultaneously while the other α/β compounded NAs could only devote to one aspect. It was summarized that the key factor of affecting the α/β compounded NAs is the crystallization temperature of iPP nucleated with NA individually (T_C). The NA with higher Tc plays a leading role in the crystallization process. Consequently the mechanical properties, crystallization properties and crystalline microstructure of iPP appear close to it. Competitive nucleation will occur when the difference of Tc between two NAs is not pronounced. According to this rule, the optimization method for compounding α and β NAs was developed. That is to find out the ratio of α and β NAs with $T_{C\ \alpha} = T_{C\ \beta}$ so as to let competitive nucleation occur during crystallization. Then the method was applied to an example of adjusting the stiffness and toughness of iPP based on different of α/β compound NAs. Rely on it the optimal ratios of α/β compounded NAs can be easily determined by calculation T_C at different ratios instead of testing them on mechanical properties. Sequentially it makes more effective to enhance stiffness and toughness of iPP based on α/β compounded NAs.

8. Acknowledgment

The authors thank the National Natural Science Funds of China (20876042), the Program of Shanghai Subject Chief Scientist (XD1401500) and the Fundamental Research Funds for the central universities of China for financial support.

9. References

Awaya, H. (1988). Morphology of different types of isotactic polypropylene spherulites crystallized from melt. *Polymer*, Vol.29, No.3, (April 1988), pp. 591-596, ISSN 00323861

Bai, H. W.; Wang, Y. & Liu, L . et al. (2008). Nonisothermal Crystallization Behaviors of Polypropylene with α/β Nucleating Agents. *J. Polym. Sci. Part B: Polym. Phys.*, Vol.17, No.46, (September 2008), pp. 1853-1867, ISSN 08876266

Bai, H. W. & Wang, Y. (2009). A comparative study of polypropylene nucleated by individual and compounding nucleating agents. I. Melting and Isothermal Crystallization. *J. Appl. Polym. Sci.*, Vol.111, No.3, (February 2009), pp. 1624-1637, ISSN 00218995

Busse, K. ; Kressler, J. & Maier, R, D., et al. (2000). Tailoring of the α-, β-, and γ-modification in isotactic polypropylene and propene/ethene random copolymers. *Macromolecules*, Vol.33, No.23, (November 2000), pp. 8775-8780, ISSN 00249297

Cai, Z. ; Zhao, S, C. & Shen, B, X., et al. (2010). The Effect of Bicyclo[2.2.1]hept-5-ene-2,3-dicarboxylate on the Mechanical Properties and Crystallization Behaviors of Isotactic Polypropylene. *J. Appl. Polym. Sci.*, Vol.116, No.2, (April 2010), pp. 792-800, ISSN 00218995

Fillon, B. ; Thierry, A. & Wittmann, J, C., et al. (1993). Self-nucleation and recrystallization of polymers. Isotactic polypropylene, β phase: β-α conversion and β-α growth transitions. *J. Polym. Sci., Part B: Polym. Phys.*, Vol.31, No.10, (September 1993), pp. 1407-1425, ISSN 08876266

Gui, Q. D. ; Xin, Z. & Zhu, W. P., et al. (2003). Effects of an organic phosphorus nucleating agent on crystallization behaviors and mechanical properties of polypropylene. *J. Appl. Polym. Sci.*, Vol.88, No.2, (April 2003), pp. 297-301, ISSN 00218995

Huang, Y. P. ; Chen, G. M. & Yao, Z., et al. (2005). Non-isothermal crystallization behavior of polypropylene with nucleating agents and nano-calcium carbonate. *Eur. Polym. J.*, Vol.41, No.11, (November 2005), pp. 2753-2760, ISSN 00143057

Ismail, A. & Al-Raheil. (1998). Isotactic polypropylene crystallization from the melt.1. Morphological study. *J. Appl. Polym. Sci.*, Vol.67, No.7, (February 1998), pp. 1259-1273, ISSN 00218995

Kristiansen, M. ; Werner, M. & Tervoort, T., et al. (2003). The binary system isotactic polypropylene/bis (3, 4-dimethylbenzylidene) sorbitol: phase behavior, nucleation, and optical properties. *Macromolecules*, Vol.36, No.14, (July 2003), pp. 5150-5156, ISSN 00249297

Labour, T. ; Ferry, L. & Gauthier, C., et al. (1999). α- and β-crystalline forms of isotactic polypropylene investigated by nanoindentation. *J. Appl. Polym. Sci.*, Vol.74, No.1, (January 1999), pp. 195-200, ISSN 00218995

Lotz, B. ; Wittmann, J, C. & Lovinger, A, J. (1996). Structure and morphology of poly (propylenes): a molecular analysis. *Polymer*, Vol.37, No.22, (October 1996), pp. 4979-4992, ISSN 00323861

Lotz, B. ; Graff, S. & Wittmann, J, C. (1986). Crystal morphology of the γ (triclinic) phase of isotactic polypropylene and its relation to the α phase. *J Polym Sci, Part B: Polym Phys,* Vol.24, No.9, (September 1986), pp. 2017-2032, ISSN 08876266

Meille, S. V. ; Bruckner, S. & Porzio, W. (1990). γ-isotactic polypropylene. A structure with nonparallel chain axes. *Macromolecules*, Vol.23, No.18, (September 1990), pp. 4114-4121, ISSN 00249297

Romankiewicz, A. ; Tomasz, S. & Brostow, W. (2004). Structural characterization of α-and β-nucleated isotactic polypropylene. *Polym. Int.*, Vol.53, No.12, (December 2004), pp. 2086-2091, ISSN 09598103

Shi, Y. Q. & Xin, Z. (2011). Study on the rule of adjustment of stiffness and toughness of isotactic polypropylene based on α/β compound nucleating agents. *Petrochemical Technology*, Vol.40, No.6, (June 2011), pp. 608-613, ISSN 10008144

Tenma, M. & Yamaguchi, M. (2007). Structure and properties of injection-molded polypropylene with sorbitol-based clarifier. *Polym. Eng. Sci.*, Vol.47, No.9, (September 2007), pp. 1441-1446, ISSN 00323888

Tordjeman, P. ; Robert, C. & Martin, G. (2001). The effect of alpha, beta crystalline structure on the mechanical properties of polypropylene. *Eur. Phys. J. E.*, Vol.4, No.4, (April 2001), pp. 495-465, ISSN 12928941

Vagar, J. (1986). Melting memory of the β-modification of polypropylene. *J. Therm. Anal.*, Vol.31, No.1, (Feb 1986), pp. 165-172, ISSN 03684466

Vagar, J. (1992). Supermolecular structure of isotactic polypropylene. *J. Mater. Sci.*, Vol.27, No.10, (Oct 1992), pp. 2557-2579, ISSN 00222461

Xu, N. (2010). Master Dissertation, East China University of Science and Technology, China

Xu, N. ; Zhao, S. C. & Xin, Z. (2011). The effect of compounded system of carboxylate β nucleating agent and α nucleating agents on nucleation and crystallization behavior of isotactic polypropylene. *Polym. Mater. Sci. Tech,* Vol.27, No.7, (July 2011), pp. ISSN 10007555

Zhang, G. P. ; Xin, Z. & Yu, J. Y., et al. (2003). Nucleating efficiency of organic phosphates in polypropylene. *J. Macromol. Sci. Part B: Phys.*, Vol.42, No.3, (July 2003), pp. 467-478, ISSN 00222348

Zhang, Y. F. & Xin, Z. (2006). Effects of substituted aromatic heterocyclic phosphate salts on properties, crystallization and melting behaviors of isotactic polypropylene. *J. Appl. Polym. Sci.*, Vol.100, No.6, (June 2006), pp. 467-478, ISSN 00218995

Zhang, Y. F. & Xin, Z. (2007). Isothermal crystallization behaviors of isotactic polypropylene nucleated with compounding nucleating agents. *J. Polym. Sci. Part B: Polym. Phys.*, Vol.45, No.5, (March 2007), pp. 590-596, ISSN 08876266

Zhao, S. C. ; Cai, Z. & Xin, Z. (2008). A highly active novel β-nucleating agent for isotactic polypropylene. *Polymer*, Vol.49, No.11, (May 2008), pp. 2745-2754, ISSN 00323861

Zhao, S. C. & Xin, Z. (2009). Crystallization Kinetics of Isotactic Polypropylene Nucleated with Organic Dicarboxylic Acid Salts. *J. Appl. Polym. Sci.*, Vol.112, No.3, (May 2009), pp. 1471-1480 ISSN 00218995

Zhao, S. C. & Xin, Z. (2010). Nucleation Characteristics of the α/β Compounded Nucleating Agents and Their Influences on Crystallization Behavior and Mechanical Properties of Isotactic Polypropylene. *J Polym. Sci.: Part B: Polym Phys,* Vol.48, No.6, (March 2010), pp. 653–665, ISSN 08876266

7

Influence of Irradiation on Mechanical Properties of Materials

V.V. Krasil'nikov[1] and S.E. Savotchenko[2]
[1]*Belgorod State University*
[2]*Belgorod Regional Institute of Postgraduate Education and Professional Retraining of Specialists*
Russia

1. Introduction

The one of actual problems of radiation material science is to reveal plastic deformation laws, hardening and fracture ones of materials under intense external action, particularly irradiation. Herein we imply different kinds of irradiation, for instance, such a (e, γ) beam irradiation, ion or neutron irradiation and so on. Evolution of a construction material microstructure at a high temperature trial operation is substantially conditioned by free migrating defects [1]. The processes of interaction of point defects with each other, with dislocations and interface underlie all of metal radiation hardening mechanisms [2]. In the section 1, the nonlinear model of dose dependence saturation of the yield strength is proposed on the base of the vacancy and interstitial barrier interaction. Processes of mutual recombination of vacancy and interstitial barriers and formation of vacancy and interstitial clusters are taken into consideration.

A series of different radiation defects (retardation barriers of dislocations), and their sizes, and a form of their volume distribution contribute into a yield strength increment for all of sorts of irradiation. The contribution of a barrier type is determined by conditions of irradiation and tests. At the low temperature irradiation (at the test temperatures up to 0.3 T_m, T_m is melting temperature), interstitial atoms, and vacancies, and their clusters contribute mainly into the hardening. In the section 2, evolution of radiation barrier (vacancies and interstitials) clusters is analyzed under low temperature radiation in the presence of the most important secondary effectes: recombination and formation of divacancy complexes. It is proposed a barrier hardening model in that mechanisms of mutual annihilation of the vacancy and interstitial barriers and their clusterization play a main role.

In the section 3 unlike two preceding sections where the dose dependences are considered, the phenomenological model is formulated to describe a yield strength temperature dependence of polycrystalline materials that have undergone irradiation and mechanical experiences in a wide temperature interval including structure levels of plastic deformation. In this section, a new phenomenological model is proposed to give a suitable description of yield strength temperature dependence of some of irradiated materials in a temperature interval including plastic deformation structure levels.

2. Mutual recombination and clusterization effect of the vacancy and interstitial barriers on radiation hardening materials

2.1 Formulation of the model

In initial stages of low temperature irradiation (up to $0.3 \cdot T_m$ where T_m is melting temperature) at small doses, inhibiting processes of initial dislocations and their sources dominate. The point defects are connect with a dislocation and form dislocation jogs and steps but interacting with each other form interstitial - vacancy - impurity clusters. As a result, energy and geometry dislocation characteristics change substantially. A phenomenological description of these mechanisms is within the usual Granato - Lukke theory.

At large irradiation doses of metals, the processes of an elastic and contact interaction of sliding dislocations with different potential barriers begin to play a main role. Besides the separated point defects in irradiated material, the dislocations are to surmount interstitial and vacancy clusters and dislocation loops, the interstitial - vacancy - impurity clusters, precipitates, voids. The dislocation can cut a barrier, can be bent by the barrier and can go round an obstacle by a dislocation climb subject to barrier intensity and a distance between barriers. These mechanisms of barrier hardening are described by the Orowan model of athermic surmounting obstacles by dislocations [3].

At the large irradiation doses, a yield strength increment can be represented by the sum of barrier contributions of different types [3]:

$$\Delta\sigma = \sum_{i=1}^{N} \Delta\sigma_i \,, \tag{2.1}$$

where index i is a barrier type, N is a number of barrier types affecting the yield strength, $\Delta\sigma_i$ is the yield strength increment of i' barrier type.

Up to now, there is many of experimental data and numerical theoretical models are developed to describe a dependence of barrier concentration of different types on radiation hardening power and behavior pattern of point defect clusters [4]. It is known that on earlier stage at small irradiation doses (by neutron radiation up to $2 \cdot 10^{16}$ n/cm^2), the hardening occurs due to forming the slowly increasing interstitial clusters. The vacancy clusters begin to contribute to the hardening in increasing dose.

In these stages, the yield strength increment is described with sufficient approximation of the dependence of the form

$$\Delta\sigma = a(\Phi t)^n \,, \tag{2.2}$$

where a is a parameter depending on irradiation conditions and the research material type, Φ is density of particle flux, t is irradiation time, exponent n changes against the material type and the irradiation condition from 0.25 to 0.75 [3].

A saturation nature of the hardening is still not clear finally. Probable causes of the yield strength increment saturation can be such as overlapping stresses field created by radiation defects of certain their concentration, creation round the volume defects of defect-free zones, the beginning of the dislocation channeling and surmounting obstacles processes and so on.

In the Ref [5], the model is proposed to describe the dose dependence of the copper yield strength increment where the saturation is explained by a decreasing velocity of the forming clusters with increasing irradiate dose due to interaction between the available clusters and newly forming ones.

Here the model is proposed to describe the dose dependence of the yield strength increment taking into account of vacancy and interstitial barrier interaction.

We consider that vacancies and interstitial atoms make a main contribution to the yield strength increment of a certain material at some of irradiation conditions. They are barriers to play the main role in the hardening at the low temperature irradiation. Therefore, in the proposed model N=2; the index values of i=1 and 2 correspond to the vacancies and interstitial atoms (and their clusters). Then in this model Eq (2.1) takes the form:

$$\Delta\sigma = \Delta\sigma_1 + \Delta\sigma_2 . \quad (2.3)$$

For all of the obstacle types, the metal yield strength increment conditioned by dislocation deceleration is described as [2]:

$$\Delta\sigma_i = \alpha_i \mu b (C_i d_i)^{1/2} , i = 1, 2, \quad (2.4)$$

where α_i is the parameters characterizing i' barrier intensity (a fixed quantity for some of barrier types, material and irradiation condition), μ is the shear modulus, b the Burgers vector length, C_i the volume density of i' barrier type, d_i their average size. For instance, the vacancy and interstitial have the average size ~ 10 nm, and the parameters characterizing barrier intensity has the value about 0.2 [2].

The present model is based on the system equations for the volume densities of the radiation- induced nonequilibrium vacancies and interstitial barriers C_1, C_2:

$$\begin{cases} \dfrac{dC_1}{d\tau} = K_1 - \dfrac{C_1}{\tau_1} - \gamma_{12} C_1 C_2 - \gamma_1 C_1^2, \\ \dfrac{dC_2}{d\tau} = K_2 - \dfrac{C_2}{\tau_2} - \gamma_{12} C_1 C_2 - \gamma_2 C_2^2. \end{cases} \quad (2.5)$$

where $\tau = \Phi t$, Φ is density of particle flux, t irradiation time, K_i, $i = 1, 2$, the intensities of forming the radiation - induced vacancy and interstitial barriers, γ_i are the coefficients of barrier recombination and characterize forming the clusters of acceptable barrier type (it can be named as clusterization coefficients), γ_{12} the coefficient of mutual recombination of the annihilating vacancy and interstitial barriers, the coefficients τ_i^{-1} can be represented by the form: $\tau_i^{-1} = K_i V_i$, where V_i are the effective volumes of interaction of the certain barriers with each other.

The first terms of the equation system (2.5) describe the intensity of increasing the volume barrier densities of the acceptable type, the second ones correspond to decreasing the volume barrier densities due to absorbing the barriers on natural sinks: voids, dislocations, dislocation network, grain boundaries and so on. In the proposed model, the mechanisms of the mutual annihilation of vacancy and interstitial barriers and their clusterization are assigned. The third terms of the system (2.5) describe decreasing the barrier densities due to of the mutual annihilation of two different type barriers, and the fourth ones do due to forming the clusters of two barriers of the same type.

To find the volume densities of the radiation - induced nonequilibrium vacancies and interstitial atoms it is necessary to set up their the initial values:

$$C_i(0) = C_i^{(0)}, \; i = 1, 2. \tag{2.6}$$

Thus, the mathematical formulation of the present model is the Cauchy problem for the system of nonlinear differential equations (2.5). The volume barrier densities found as a result of solution of the Cauchy problem (2.5), (2.6) determine the yield strength increment according to Eqs (2.3), (2.4).

2.2 Dose saturation features of the yield strength subject to annihilation effects of the vacancy and interstitial barriers

At the beginning, we consider the yield strength behavior against the material irradiation dose taking into account of only the mutual recombination of the vacancy and interstitial barriers that is their annihilation. Upon that, the effects of forming the barrier clusters of the same type are not considered. The barrier annihilation effects but no their clusterization play the main role in the radiation hardening. Therefore $\gamma_{12} >> \gamma_i$, $i = 1, 2$, and it is possible to neglect the last terms of $\gamma_i C_i^2$ in every equation of the system (2.5).

In this approximation the system (2.5) takes the form:

$$\begin{cases} \dfrac{dC_1}{d\tau} = K_1(1 - V_1 C_1) - \gamma_{12} C_1 C_2, \\ \dfrac{dC_2}{d\tau} = K_2(1 - V_2 C_2) - \gamma_{12} C_1 C_2. \end{cases} \tag{2.7}$$

As Eqs (2.7) has not an analytic solution, we illustrate a dependence pattern of the barrier density by a numerical analysis example of the Cauchy problem solution in which the parameter simulating values and the initial defect densities $C_i^{(0)} = 5{\cdot}10^{13}\text{cm}^{-3}$ (they are pointed out in the figure captures following then) are used. The numerical solution results of the Cauchy problem (2.6), (2.7) reduced to the dimensionless form attached to the parameter simulating values are shown in Fig.2.1.

Here it is shown the dependence typical behavior of the barrier relative densities of the vacancy C_1 / C_0 and interstitial C_2 / C_0 types on the relative irradiate dose τ/τ_0 where C_0 is a measure scale of the defect densities (here it is equal 10^{15} cm^{-3}). The measure units are accepted relative to the selected scale τ_0 (fluence) the numerical value of which is determined by a specific problem (it is convenient to select a minimal fluence of the specific problem as the dose measure scale; for instance, in the ion irradiation the value of τ_0 can be equal 10^{14} ion/cm^2 or 10^{22} n/m^2 as the contemporary neutron fluence). This dependence pattern is universal and independent of the selected measure scales of the specific physics parameters. In increasing dose it is happened the saturation of the material by the radiation - induced barriers (their densities do not increased far more).

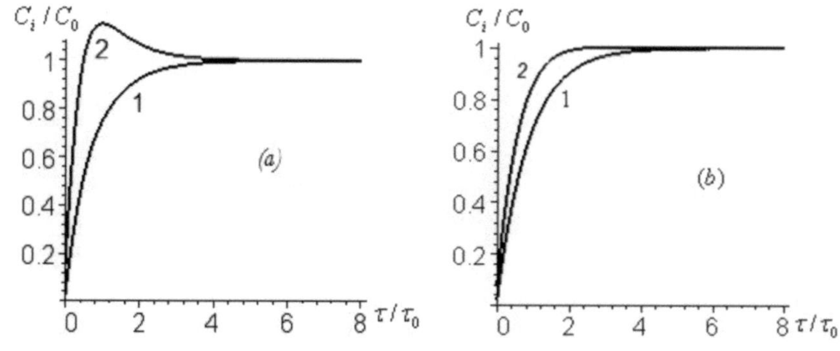

Fig. 2.1. Dependences of relative barrier densities of vacancy type (the curve 1) and interstitial type (the curve 2) on irradiate dose at fixed values of the dimensionless parameters $(K_1\tau_0)/C_0 = 2.5$, $\tau_0/\tau_1 = 1.1$, $(K_2\tau_0)/C_0 = 1$, $\tau_0/\tau_2 = 1.5$, and different values of the recombination coefficient: (*a*) – $\gamma_{12}C_0\tau_0 = 0.9$, (*b*) – $\gamma_{12}C_0\tau_0 = 0.1$ where C_0, τ_0 are the measure scales of barriers and dose (fluence) respectively the concrete selection of that is conditioned by the concrete problem.

The numerical analysis shows that at $K_1 > K_2$ and small doses, the barrier density of the vacancy type exceeds the one of the interstitial type. At opposite inequality, a situation becomes reverse that is natural as it physically means that increasing the irradiation intensity leads to enlarging a number both the vacancy and the interstitial barriers. Upon that the higher the velocity of forming the radiation barriers of any type is the larger their volume density. It is necessary to mark that the vacancy barrier density usually is higher than the interstitial barrier density in the real materials near a sample surface irradiated.

The numerical analysis also shows that increasing the mutual recombination coefficient γ_{12} leads to decreasing the saturation values of the barrier densities. It leads as well to changing the dose dependence pattern of the barrier density of the corresponding type that becomes the monotonic quantity (see Fig.2.1, a) and b)).

Substitution of the found numerical solution of the system (2.7) to Eq (2.3) allows getting the dose dependences of the material yield strength increment. For plotting these dependence figures, it is convenient to represent Eq (2.3) as follows

$$\Delta\sigma = \Delta\sigma_\infty^0\{(d_1C_1 / dC_0)^{1/2} + (d_2C_2 / dC_0)^{1/2}\}, \qquad (2.8)$$

$$\Delta\sigma_\infty^0 = \alpha\mu b(C_0 d)^{1/2} \qquad (2.9)$$

where d is an average size of the barrier clusters of all types (here it can be determined as half-sum of d_1 and d_2).

The results of numerical modeling the behavior of the yield strength are represented in Fig. 2.2. where the typical dose dependences of the yield strength increment are shown on the base of the obtained Eqs (2.8) and (2.9) (in relative units) at the model parameter values. The corresponding numerical analysis shows that the yield strength gets the saturation quickly enough. The typical monotonic form of the dose saturation plots of the yield strength does not change virtually in a broad enough interval of the model parameter values satisfying to the existence condition of the Cauchy problem (2.5), (2.6) solution.

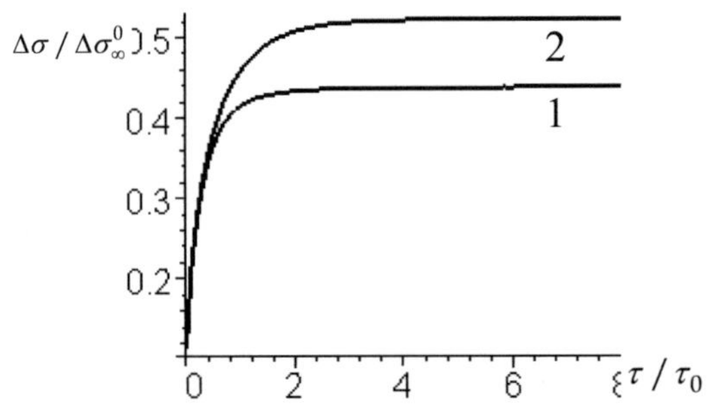

Fig. 2.2. Dependences of the relative yield strength increment on irradiate dose at fixed the nondimensional parameters $(K_1\tau_0)/C_0 = 2.5$, $\tau_0/\tau_1 = 1.1$, $(K_2\tau_0)/C_0 = 1$, $\tau_0/\tau_2 = 1.5$, and different values of the recombination coefficient: (1) $\gamma_{12}C_0\tau_0 = 0.9$, (2) $\gamma_{12}C_0\tau_0 = 0.1$ in relative units.

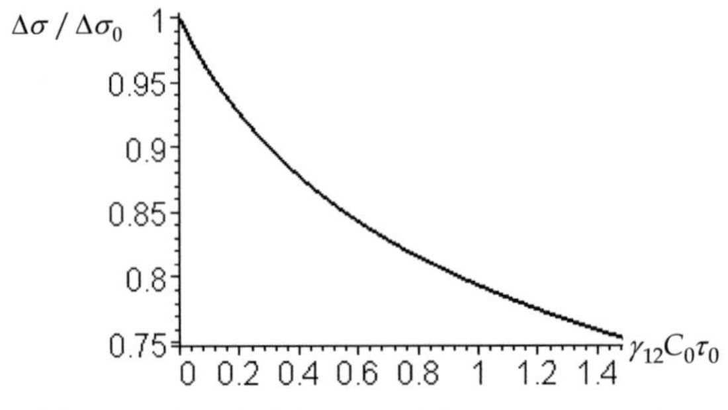

Fig. 2.3. Dependences of the relative yield strength increment on the recombination coefficient at fixed irradiate dose $\tau/\tau_0 = 4$ and the rest parameter values as in Fig. 2.2.

Besides, on the base of the numerical analysis, it is obtained that a growth of the mutual recombination coefficient γ_{12} leads to decreasing the saturation value of the yield strength. It is shown in Fig.2.3. where the dependence of the relative yield strength increment $\Delta\sigma / \Delta\sigma_0$ on the mutual recombination coefficient at fixed dose is shown (here $\Delta\sigma_0$ is the yield strength increment value at $\gamma_{12} = 0$ that is under the condition of total neglecting interaction between the vacancy and interstitial barriers). It is well seen that the relative yield strength increment decreases with the growth of the mutual recombination coefficient, this dependence being nonlinear and monotonic.

A stationary point of the systems (2.7) determines the saturation values of the barrier density:

$$C_1^{(s)} = \frac{K_1}{K_1 V_1 + \gamma_{12} C_2^{(s)}}, \tag{2.10}$$

$$C_2^{(s)} = C_a \left(1 + \sqrt{1 + \frac{C_b}{C_a}} \right), \tag{2.11}$$

where

$$C_a = \frac{(K_2 - K_1)\gamma_{12} - K_1 K_2 V_1 V_2}{2\gamma_{12} K_2 V_2}, \quad C_b = \sqrt{\frac{K_1 V_1}{\gamma_{12} V_2}}.$$

The saturation values of the barrier density (2.10) and (2.11) are achieved quickly enough (see Fig.2.1.).

In the case when the recombination of the vacancy and interstitial barriers is negligibly small the equations of the system (2.7) become independent at $\gamma_{12} = 0$. If to suppose in addition that the radiation barriers are virtually absent at the initial time $C_i^{(0)} = 0$, ($i = 1, 2$) then as a result the known expression of the volume barrier density is obtained [3]:

$$C_i(\tau) = \frac{1}{V_i}(1 - e^{-\tau/\tau_i}), \quad i = 1, 2, \tag{2.12}$$

where $\tau_i = (K_i V_i)^{-1}$. Substitution of Eq (2.12) to Eq (3) gives the expression of the yield strength increment at $\gamma_{12} = 0$:

$$\Delta\sigma_0 = \Delta\sigma_1^{(s)}(1 - e^{-\tau/\tau_1})^{1/2} + \Delta\sigma_2^{(s)}(1 - e^{-\tau/\tau_2})^{1/2}, \tag{2.13}$$

where $\Delta\sigma_i^{(s)} = \alpha_i \mu b(d_i / V_i)^{1/2}$, $i = 1, 2$. In this case, at large doses it is followed from Eq (2.13) that. $\Delta\sigma_0 = \Delta\sigma_1^{(s)} + \Delta\sigma_2^{(s)}$

It should be noted that at small doses it is followed the well known law of Eq (2.2) at $n = ½$:

$$\Delta\sigma_0 = a(\Phi t)^{1/2}, \tag{2.14}$$

where $a = \mu b\{\alpha_1 (K_1 d_1)^{1/2} + \alpha_2 (K_2 d_2)^{1/2}\}$.

In the absence of the mutual recombination of the vacancy and interstitial barriers but at the arbitrary initial barrier densities and the initial conditions (2.6) and at $\gamma_{12} = 0$, the solution of the equation system (2.7) leads to the expression

$$C_i(\tau) = \frac{1}{V_i}\{1 + (C_i^{(0)} V_i - 1)e^{-K_i V_i \tau}\}, \; i = 1,2. \tag{2.15}$$

Substituting Eq (2.15) into (2.3) it can be obtained the expression for the yield strength increment at $\gamma_{12}=0$ and the arbitrary initial barrier densities:

$$\Delta\sigma_0 = \Delta\sigma_1^{(s)}\{1 + (C_1^{(0)} V_1 - 1)e^{-K_1 V_1 \tau}\}^{1/2} + \Delta\sigma_2^{(s)}\{1 + (C_2^{(0)} V_2 - 1)e^{-K_2 V_2 \tau}\}^{1/2} \tag{2.16}$$

Eq (2.16) is used to plot the yield strength dependence on the intensity quantity of the vacancy and interstitial barrier recombination at fixed dose in Fig. 2.3.

2.3 Dose saturation features of the yield strength subject to the effects of clustering barriers

Let us consider now the main contribution to the radiation hardening is given by the effects of forming the clusters of two barriers of the same type and the mutual recombination of the vacancy and interstitial barriers is negligibly small. Therefore, $\gamma_{12} << \gamma_i$, $i = 1, 2$, and in every equation of the system (2.5) it can be neglected by the last terms of $\gamma_{12}C_1C_2$. After this, the system equations become independent and therefore in what follows the indexes 1 and 2 of notations can be omitted.

Further we consider the contribution to the radiation hardening only from the barriers of the same type (either vacancy or interstitial) and then instead Eqs (2.3) and (2.4) the next expression is used

$$\Delta\sigma = \alpha\mu b(Cd)^{1/2}, \tag{2.17}$$

where α is the parameter characterizing the barrier intensity (a fixed quantity for some of barrier types), C the volume barrier density of the same type, d their average size.

In this case it is convenient to go to single equation of the system (2.5):

$$\frac{dC}{d\tau} = K(1 - VC - \gamma C^2), \tag{2.18}$$

where γ_i of Eqs (2.5) is changed by the new notation according to the relationship $\gamma_i = K\gamma$ and all of the indexes are omitted.

The solution of Eq (2.18) with the initial condition C (0) = 0 at $\gamma > 0$ takes the form

$$C(\tau) = \frac{1}{2\gamma}\left\{q\,\mathrm{th}\left(\frac{\tau}{\tau'} + \varphi\right) - V\right\}, \tag{2.19}$$

where $\tau' = 2/Kq$, $q = \sqrt{V^2 + 4\gamma}$, $\varphi = \mathrm{Arth}\frac{V}{q} = \frac{1}{2}\ln\frac{q+V}{q-V}$. The found expression (2.19) describes the dependence of the volume radiation-induced barrier density on irradiate dose (fluence) $\tau = \Phi t$. The typical plot of the dependence (2.19) is shown in Fig.2.4. at the different values of barrier clusterization intensity and the fixed rest parameters in the relative units (as stated above).

Substituting Eqs (2.19) into Eqs (2.17) we obtain the dose dependence of the yield strength increment:

$$\Delta\sigma = \Delta\sigma_\infty^0 \frac{1}{\sqrt{2\gamma C_0}}\left\{q\,\mathrm{th}\left(\frac{\tau}{\tau'} + \varphi\right) - V\right\}^{1/2} \tag{2.20}$$

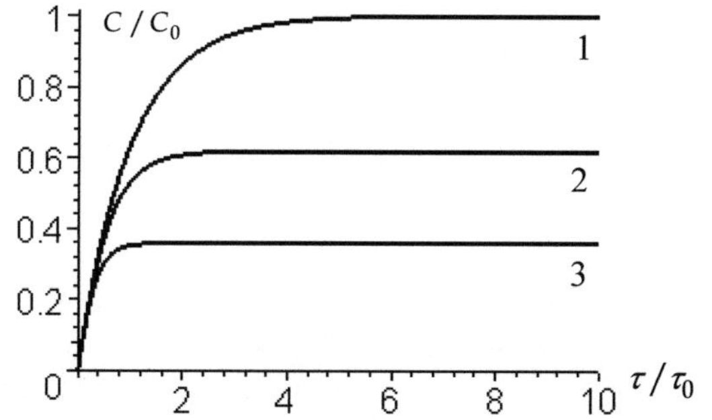

Fig. 2.4. Dose dependences of the volume barrier density (2.19) at the fixed parameter values $(K\tau_0)/C_0 = 1$, $C_0V = 1$ and the different values of clusterization intensity: (1) $C_0^2\gamma = 0$, (2) $C_0^2\gamma = 1$, (3) $C_0^2\gamma = 5$ in relative units.

In the present model, it is supposed that the average barrier size is a weakly changing function of irradiate dose. The typical plot of the dependence (2.20) is shown in Fig. 2.5 at the different values of barrier clusterization intensity and the fixed rest parameters.

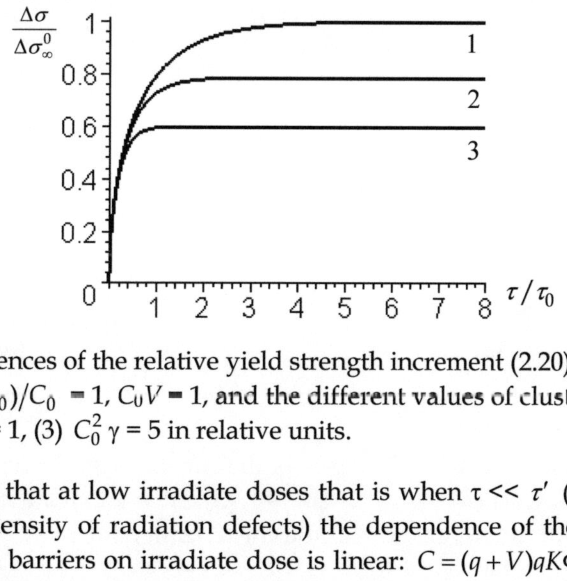

Fig. 2.5. Dose dependences of the relative yield strength increment (2.20) at the fixed parameter values $(K\tau_0)/C_0 = 1$, $C_0V = 1$, and the different values of clusterization intensity. (1) $C_0^2\gamma = 0$, (2) $C_0^2\gamma = 1$, (3) $C_0^2\gamma = 5$ in relative units.

It follows from (2.19) that at low irradiate doses that is when $\tau << \tau'$ (accordingly (2.19) it corresponds to low density of radiation defects) the dependence of the volume density of the radiation-induced barriers on irradiate dose is linear: $C = (q+V)qK\Phi t$. Substituting this expression into (2.17) we obtain the dependence of $\Delta\sigma = a(\Phi t)^{1/2}$ where $a = \alpha\mu b\{(q+V)qKd\}^{1/2}$.

At high irradiate doses that is when $\tau >> \tau'$ the volume density of the radiation-induced barriers is saturated and tends to the constant $C_\infty = (q-V)/2\gamma$. This value of the volume density is the stationary point of Eq (2.18). Substituting this expression into Eqs (2.17) we obtain the saturation value of the yield strength:

$$\Delta\sigma_\infty = \Delta\sigma_\infty^0 \left(\frac{q-V}{2\gamma C_0}\right)^{1/2} \tag{2.21}$$

where $\Delta\sigma_\infty^0$ is determined by Eq (2.9).

If the clusterization effects can be neglected (as well as the mutual annihilation of the barrier of the different types has been already neglected too) then it can be obtained from Eq (2.19) the known expression of the volume barrier density as Eq (2.12). This dependence corresponds to the curve 1 in Fig.2.4.

After substituting such volume density into Eq (2.17) it is obtained the dose dependence of the yield strength in the case when the barrier clusterization makes negligibly small contribution to the velocity of forming the radiation – induced barriers [3]:

$$\Delta\sigma^0 = \Delta\sigma_\infty^0\left(\frac{1-e^{-\tau/\bar{\tau}}}{C_0 V}\right)^{1/2}, \qquad (2.22)$$

where $\bar{\tau} = (KV)^{-1}$. This dependence corresponds to the curve 1 in Fig.2.5. Hence it follows at small irradiate doses the well known law as $\Delta\sigma = a(\Phi t)^{1/2}$ where now $a = \alpha\mu b(Kd)^{1/2}$.

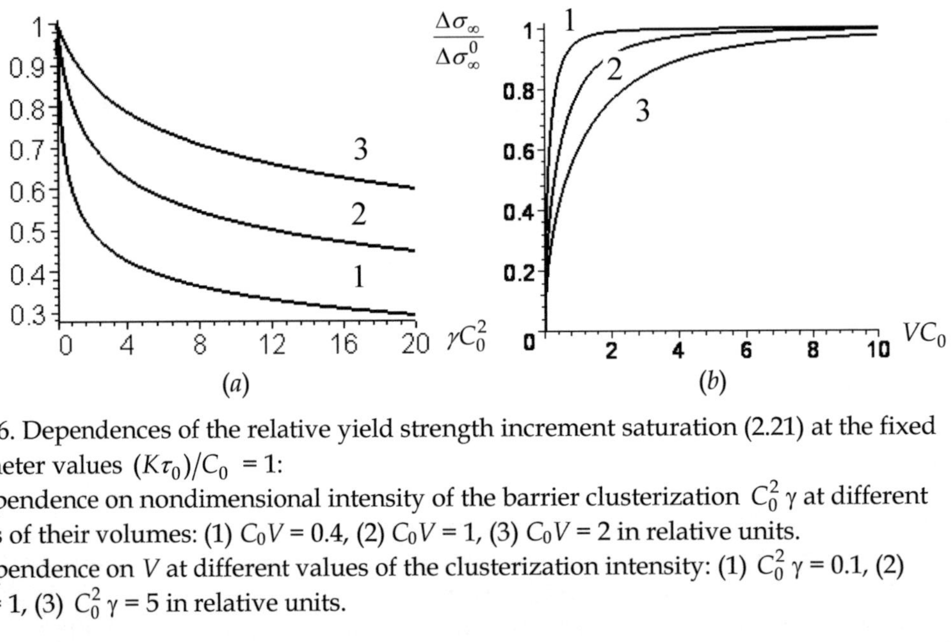

Fig. 2.6. Dependences of the relative yield strength increment saturation (2.21) at the fixed parameter values $(K\tau_0)/C_0 = 1$:
(*a*) dependence on nondimensional intensity of the barrier clusterization $C_0^2\gamma$ at different values of their volumes: (1) $C_0V = 0.4$, (2) $C_0V = 1$, (3) $C_0V = 2$ in relative units.
(*b*) dependence on V at different values of the clusterization intensity: (1) $C_0^2\gamma = 0.1$, (2) $C_0^2\gamma = 1$, (3) $C_0^2\gamma = 5$ in relative units.

On the base of the obtained dependence of the yield strength saturation by Eq (2.21) on the barrier recombination intensity, it can be drawn the next conclusion. The saturation quantity of the material yield strength at fixed dose decreases monotonically with increasing the intensity of radiation - induced barrier clusterization. This behavior is shown in Fig.2.6. where the plots of Eq (2.21) are presented.

2.4 Discussion of experiment data

As known, since 1988 the EU, USA, Japan and Russia joint works have been fulfilled within the intergovernmental agreement approved by IAEA (International Atomic Energy

Agency). According to technological design of the thermonuclear reactor ITER, one of main constructional material is austenitic steel 316(N)-IG [6].

In the Ref [9], it is developed the model of the radiation hardening of the concrete irradiation material, 316(N)-IG steel. This model is based on the equation to be equivalent to Eq (2.18) for the potential barrier density *C* the role of which plays the stacking fault tetrahedral observed by electron microscopy as black dots. Experiment reveals that the concentration of these barriers grows with increasing irradiation dose.

Authors of the work [9] make a comparison with experimental data for 316(N)-IG steel on the base of the equation analogous to above Eq (2.22) to be the particular case of Eq (2.18) that is not taking into account of the barrier annihilation. They receive relatively good fit with the experimental data in temperature ranges 20 – 150 and 230 - 300°C. For the higher temperatures (330 - 400°C) this equation does not obey an adequate description of the yield strength and ultimate stress increment. To describe the experimental data the authors of Ref [7] fit an exponent in the dependence of the form (2.17) pointing out this exponent to be varied approximately in the interval 1.4 – 2.7 for the best agreement with the experimental data. It is possible that accounting the barrier recombination of the different types leads to invariability of the exponent ½ in Eq (2.17).

In the work [8], the equation analogous to Eq (2.22) it is used for fitting the experimental data for strength and ductility of corrosion – resisting austenitic 06X18H10T steel irradiated by the WWER – 440 reactor up to damaging dose ~ 21 d.p.a. at different testing temperatures. Authors of Ref [8] find out that the radiation hardening saturation of 06X18H10T steel irradiated in WWER-440 reactors takes place at the damage dose of ~ 10 ÷ 15 d.p.a. ($1 \div 1{,}5 \cdot 10^{26}$ n/cm^2).

3. Effect of secondary processes on material hardening under low temperature radiation

3.1 Formulation of the secondary process contribution model

The base of the model is a barrier mechanism of the radiation hardening [2] according to that the yield strength increment can be represented by the sum of barrier contributions of different types (see Eq (2.1) of the preceding section).

We consider that the barriers of vacancy and interstitial types make a main contribution to the yield strength increment of a certain material. These barriers play the main role in the hardening at the low temperature irradiation. Therefore, in the proposed model N=3; the index values of i=1 correspond to the vacancy barriers, i = 2 do to interstitial ones and i = 3 do to more large vacancy complexes. Then in this model Eq (2.1) takes the form:

$$\Delta\sigma = \Delta\sigma_1 + \Delta\sigma_2 + \Delta\sigma_3 \,. \tag{3.1}$$

For all of the obstacle types, the metal yield strength increment conditioned by dislocation deceleration is described by the Eq (2.4) (see the preceding section).

Under irradiation, development of radiation defect clusters (barriers) of different types occurs in a region of a primary knocked-on atom. It is proposed that the interstitial barriers have considerably smaller sizes and leave the damage region of a sample sooner than the vacancy barriers do. In connection with this, we formulate the phenomenological model that is based on the equation system for volume densities of the radiation-induced non-equilibrium vacancy C_1 and interstitial barriers C_2 and more large complexes C_3 developed by bimolecular mechanism of the vacancy barriers:

$$\begin{cases} \dfrac{\partial C_1}{\partial \tau} = D_1 \Delta C_1 + K_1 - C_1 / \tau_1 - \gamma_{12} C_1 C_2 - \gamma_1 C_1^2, \\ \dfrac{\partial C_2}{\partial \tau} = D_2 \Delta C_2 + K_2 - C_2 / \tau_2 - \gamma_{12} C_1 C_2 - \gamma_2 C_2^2, \\ \dfrac{\partial C_3}{\partial \tau} = D_3 \Delta C_3 + \gamma_1 C_1^2 - C_3 / \tau_3. \end{cases} \tag{3.2}$$

Here $\tau = \Phi t$, Φ is the particle flux, t irradiation time, K_i, $i = 1, 2$, the intensities of forming the radiation - induced vacancy and interstitial barriers, γ_i are the coefficients of barrier recombination and characterize forming the clusters of acceptable barrier type (it can be named as clusterization coefficients), the coefficients τ_i^{-1} can be represented by the form: $\tau_i^{-1} = K_i V_i$, where V_i are the effective volumes of interaction of the certain barriers with each other, γ_{12} the coefficient of mutual recombination of the annihilating vacancy and interstitial barriers. It can be valued as follows:

$$\gamma_{12} = \frac{4\pi r (D_1 + D_2)}{\Omega} e^{-\frac{E_r}{k_B T}},$$

where Ω atom volume, E_r activation energy of recombination of vacancy and interstitial barriers, r recombination radius, T test temperature, k_B Boltzmann constant, D_i diffusion coefficients of the non-equilibrium barriers of the given types: $D_i = D_{i0} \exp(-E_i / k_B T)$, $i = 1, 2$, where E_i energy of activation and migration of respective barriers, $D_{i0} = a^2 \nu$, a and ν are length and barrier jumping frequency for migration, respectively.

As material structure changes go under irradiation for times large in comparison with relaxation time of point defects then only diffusion barrier processes are considered to be very slow and therefore we neglect diffusion terms in the equations of the system (3.2). In addition, we study evolution of barrier volume densities in time considering their distributions are spatially homogeneous. In this case, the system (3.2) takes the form:

$$\begin{cases} \dfrac{\partial C_1}{\partial \tau} = K_1 - C_1 / \tau_1 - \gamma_{12} C_1 C_2 - \gamma_1 C_1^2, \\ \dfrac{\partial C_2}{\partial \tau} = K_2 - C_2 / \tau_2 - \gamma_{12} C_1 C_2 - \gamma_2 C_2^2, \\ \dfrac{\partial C_3}{\partial \tau} = \gamma_1 C_1^2 - C_3 / \tau_3. \end{cases} \tag{3.3}$$

In this equation system, the first two equations coincide with the system (2.5) completely (see the preceding section).

The third equation of the system (3.3) describes redistribution of divacancy barrier complexes. Contribution to the hardening due to divacancy barriers is determined by only vacancy barrier density. The kinetic coefficients τ_3^{-1} and γ_1 characterize intensities of breakdown of vacancy clusters and development of divacancy clusters. Effect of these contributions are appreciable if intensity of secondary processes of developing divacancy complexes predominates over their breakdown.

To find the volume densities of the nonequilibrium barriers it is necessary to set up their initial values:

$$C_i(0) = C_i^{(0)},\ i = 1, 2, 3 \tag{3.4}$$

Thus, the mathematical formulation of the model proposed in this section is the Cauchy problem for the system of nonlinear differential equations (3.3) with initial conditions (3.4). The volume barrier densities found as a result of solution of the Cauchy problem (3.3), (3.4) are to be inserted into Eq (2.4) that determines the total yield strength increment (3.1).

3.2 Numerical analysis of the model results

The model values of parameters (are given in captures of Figures) and the initial conditions $C_i^{(0)} = 5{\cdot}10^{13}\ cm^{-3}$ .are used to fulfill numerical analysis of the Cauchy problem (3.3), (3.4). The results of the numerical solution of the Cauchy problem (3.3), (3.4) (reduced to the dimensionless form) are represented on Fig.3.1. at the indicated parameter values.

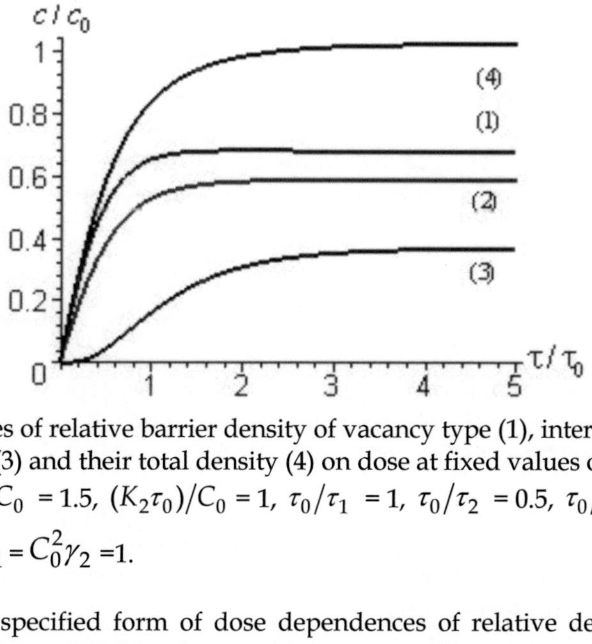

Fig. 3.1. Dependences of relative barrier density of vacancy type (1), interstitial type (2), vacancy complexes (3) and their total density (4) on dose at fixed values of dimensionless parameters $(K_1\tau_0)/C_0 = 1.5$, $(K_2\tau_0)/C_0 = 1$, $\tau_0/\tau_1 = 1$, $\tau_0/\tau_2 = 0.5$, $\tau_0/\tau_3 = 1.25$, $\gamma_{12}C_0\tau_0 = 0.9$, $C_0^2\gamma_1 = C_0^2\gamma_2 = 1$.

Here is shown the specified form of dose dependences of relative densities for vacancv barriers C_1 / C_0, and interstitial barriers C_2 / C_0, and more large vacancy complexes C_3 / C_0 where C_0 is a measure scale of barrier density taken to be equal $10^{15}\ cm^{-3}$ in this case on dose τ/τ_0 in units measured by the scale τ_0 (it is convenient to select a minimal fluence of the specific problem as the dose measure scale; for instance, in the ion irradiation the value of τ_0 can be equal 10^{14} ion/cm^2 or 10^{22} n/m^2 in the neutron irradiation and so on, for the specific problem, respectively). These dependence patterns are universal and independent of the selected measure scales of the specific physics parameters. It is shown that the saturation of the material by the radiation - induced barriers takes place with increasing dose.

The numerical solution of the Cauchy problem (3.3), (3.4) permits to obtain the dose dependences of the total increment of material yield strength which is convenient to represent for construction of graph as follows

$$\Delta\sigma = \Delta\sigma_{\infty}^{0}\sum_{i=1}^{3}\sqrt{d_i C_i / dC_0}\ , \qquad (3.5)$$

where $\Delta\sigma_{\infty}^{0}$ is determined by Eq (2.9) (see the preceding section), d average size of barrier cluster over all of types. The results of numerical modeling the behavior of the yield strength increment are represented in Fig. 3.2.

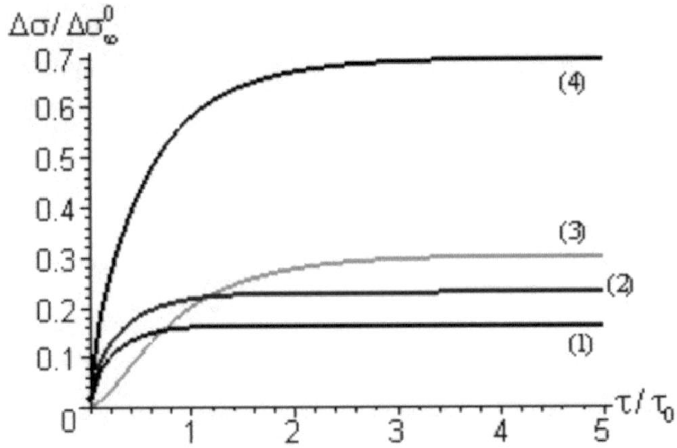

Fig. 3.2. Dose dependences of barrier contributions of vacancy type (1), interstitial type (2), vacancy complexes (3) and their total density (4) to yield strength increment at fixed values of parameters the same as in Fig.3.1. and $d_1/d = 0.016$, $d_2/d = 0.035$, $d_3/d = 0.097$.

The numerical analysis shows that the yield strength increment gets the saturation quickly enough. The typical monotonic form of the dose saturation plots of the yield strength increment does not change virtually in a broad enough interval of the model parameter values satisfying to the existence condition of the Cauchy problem (3.3), (3.4) solution.

It is should be noted though the vacancy complexes have lower concentration in comparison with vacancies and interstitial atoms they, due to their larger sizes, contribute more considerably to yield strength increment at dose build-up.

3.3 Secondary reaction contribution analysis

Let us consider the case when secondary reactions play a main role that is barrier recombination goes less intensively than developing barrier clusters. In this extreme case, we consider $\gamma_{12} \ll \gamma_1$ and $\gamma_{12} \ll \gamma_2$. Then the second equation of the system (3.3) becomes independent and coinciding formally with the first equation. In the result, the system (3.3) consists of two equations:

$$\begin{cases} \dfrac{\partial C_1}{\partial \tau} = K_1 - C_1 / \tau_1 - \gamma_1 C_1^2, \\ \dfrac{\partial C_3}{\partial \tau} = \gamma_1 C_1^2 - C_3 / \tau_3. \end{cases} \tag{3.6}$$

The first equation of the system (3.6) doesn't contain C_3. Therefore, it is independent. Its solution with zero initial condition takes the form:

$$C_1(\tau) = C_a \tanh\left(\frac{\tau}{\tau_c} + \varphi\right) - C_b , \tag{3.7}$$

where $\tau_c = 2\tau_1/\kappa$, $C_a = \kappa/2\tau_1\gamma_1$, $C_b = C_a/\kappa$, $\kappa = \sqrt{1 + 4K_1\gamma_1\tau_1^2}$, $\varphi = \mathrm{Artanh}(1/\kappa)$. The obtained expression (3.7) describes the dependence of vacancy barrier volume density on dose $\tau = \Phi t$.

When the processes of vacancy barrier clusterization are absent overall ($\gamma_1 = 0$) it results from (3.6) $C_1(\tau) = K_1\tau_1(1 - e^{-\tau/\tau_1})$ whence the well known contribution to yield strength increment follows in the case of hardening by the barrier of a single type:

$$\Delta\sigma_1 = \Delta\sigma_\infty^0 \{d_1 K_1 \tau_1 (1 - e^{-\tau/\tau_1}) / dC_0\}^{1/2} . \tag{3.8}$$

Substituting Eq (3.7) into the second equation of the system (3.6) its solution can be written as

$$C_3(\tau) = \gamma_1 \int_0^\tau e^{(s-\tau)/\tau_3} C_1^2(\tau) ds . \tag{3.9}$$

Further, this expression is used to analysis the vacancy complex contribution to yield strength increment:

$$\Delta\sigma_3 = \Delta\sigma_\infty^0 (d_3 C_3 / dC_0)^{1/2} . \tag{3.10}$$

The increasing of saturation quantity of yield strength increment goes with increasing intensity of the clusterization processes that is with increasing parameter γ_1 (Fig.3.3*a*).

This increasing is nonlinear, enlargement of yield strength increment saturation going less and less considerably with increasing intensity of the clusterization processes.

Growth of saturation quantity of yield strength increment goes with increasing specific times τ_1 and τ_3 as well (Fig.3.3. *b* and *c*).

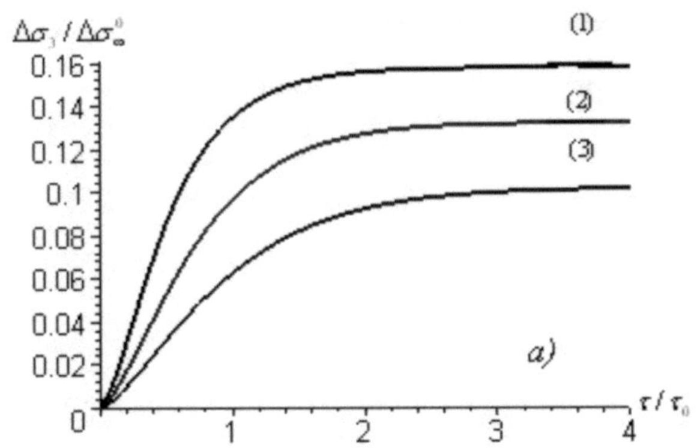

Fig. 3.3. *a*. Dose dependences of vacancy complex contribution to yield strength increment at fixed values of parameters $(K_1\tau_0)/C_0 = 1$, $d_3/d = 0.097$, $\tau_1/\tau_0 = 1$, $\tau_3/\tau_0 = 0.4$, (1) – $C_0^2\gamma_1 = 5$; (2) – $C_0^2\gamma_1 = 1.5$; (3) – $C_0^2\gamma_1 = 0.5$.

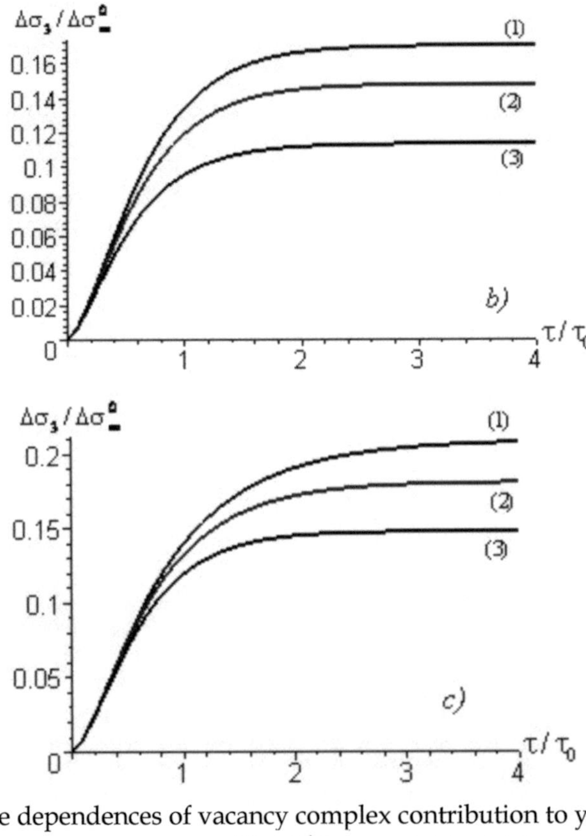

Fig. 3.3. *b* and *c*. Dose dependences of vacancy complex contribution to yield strength increment at fixed values of parameters $(K_1\tau_0)/C_0 = 1$, $d_3/d = 0.097$,

b) $\tau_3/\tau_0 = 0.4$, $C_0^2\gamma_1 = 3$, (1) – $\tau_1/\tau_0 = 2$; (2) – $\tau_1/\tau_0 = 1$; (3) – $\tau_1/\tau_0 = 0.5$;

c) $\tau_1/\tau_0 = 1$, $C_0^2\gamma_1 = 3$, (1) – $\tau_3/\tau_0 = 0.8$; (2) – $\tau_3/\tau_0 = 0.6$; (3) – $\tau_3/\tau_0 = 0.4$.

4. Phenomenological model of yield strength dependence on the temperature of irradiated materials

4.1 Temperature intervals of radiation embrittlement with taking into account two components of material flow stress

The results of experimental studying radiation embrittlement effects and the temperature dependences of such durable material characteristics as specific elongation and yield strength have been given in a series of the works [10-15]. In Refs [12, 15], it is shown that a deformation process connected with dislocation collective behavior in irradiated deformed materials is characterized by availability of the different structure deformation levels.

As known under irradiation material plastic properties undergo strong changes. In particular a radiation embrittlement phenomenon takes place [11]. Upon that plastic properties of irradiated materials depend essentially on temperature. It is interesting to analyze phenomena of radiation embrittlement and radiation hardening of reactor materials with taking into account their durable characteristics on temperature.

To analyze radiation embrittlement it is necessary to take into account availability of two components of material flow stress (σ): the thermal (thermo activated) component σ^* created by short - rang forces and the athermal one σ_μ determined by long - range forces of slowing - down dislocations and no experiencing influence of temperature. These components are shown on the plot of a temperature dependence of material flow stress (Fig.4.1.).

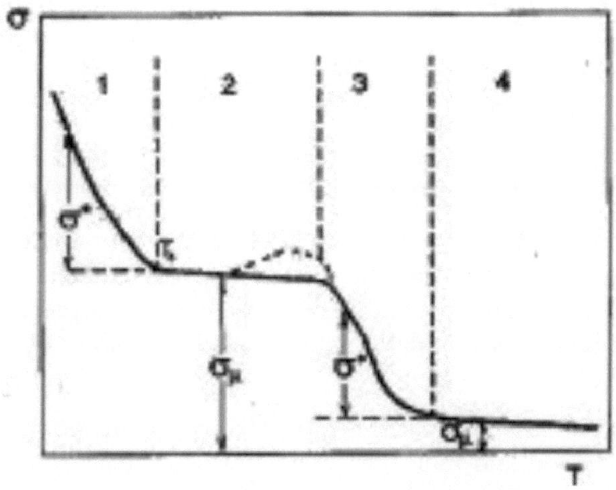

Fig. 4.1. Generalized scheme of temperature dependence of flow stress in polycrystalline materials. Area 1 corresponds to low temperature range T<0,15 T_m; area 2 is characterized by athermal component σ_μ up to ~0,45 T_m; area 3 corresponds to thermo activated component of flow stress; area 4 corresponds to second athermal plateau σ_μ ~G.

The first activated area (1) on Fig. 4.1. covers a low temperature interval $T \le 0{,}15 \cdot T_m$ in which a activative volume quantity of plastic deformation is as b^3 where b is the Burgers vector modulus. This corresponds to a microscale level of dislocation interactions that is realized by point kinetics processes of dislocations. The area (2) on Fig. 4.1. is characterized by availability of athermal component σ_μ which is mainly determined by long - range elastic internal stresses forming due to interaction of dislocations moving in parallel or crossing sliding planes. At temperatures of $T \geq 0{,}45 \cdot T_m$ (the area (3) of Fig. 4.1.), edge dislocation creeping conditioned by diffusion processes and forming crew dislocation jogs are determined by the thermo activated flow stress component. In the area (4) of Fig. 4.1., intensification of grain boundary processes of plastic deformation takes place and forms a second athermal plateau $\sigma_\mu \sim G$.

Connection of σ^* and σ_μ changes with temperature dependence of radiation embrittlement in a wide enough experience temperature interval including certain areas shown on Fig. 4.1. can be studied by a method of modeling neutron irradiation action by relativistic electron beams with energies exceeding nuclear reaction threshold (so called $((e, \gamma)$ - beams). Such irradiation as well as reactor one leads to forming different radiation defects (for instance, defects of diclocation loop type) besides nuclear reaction products) [3, 16].

Main preference of such beams is a possibility to create for short time (for some of hours) radiation damages equivalent to ones obtained for some of years of irradiation in reactors. Besides modeling experiments can be fulfilled under severely controlled conditions that has paramount importance to clear up mechanisms of phenomena in nuclear and thermonuclear reactor materials under exploitation.

The beams of electrons and γ - quanta having a large track length in materials make it possible to create homogeneous radiation damages in samples assigned for investigating mechanical properties. Investigations of mechanical prosperities of materials irradiated by (e,γ) - beams showed availability of their employment for modeling reactor damages and selection of construction material [17].

When high energy electrons get through substance an electromagnetic avalanche develops. In increasing electron penetration depth in to a material sample a number of avalanche particles increases, energy of electron decreases and the X-ray bremsstrahlung increases.

Irradiation of materials by high energy electrons leads to accumulation of large amount of helium due to secondary (γ,α) - reactions, which is accountable for high temperature radiation embrittlement. The unique feature of (e,γ) - beams is a possible to receive samples with distinct ratio of helium accumulation rate to rate of forming displacements at the same experiment.

Changing elastic and inelastic properties of polycrystalline materials are caused by lattice damages under irradiation and their next interaction with dislocations. Diffusion of point defects plays important role in process of pinning dislocations in connection with that it can be obtained significant information about radiation defects investigating influence of experience temperature on a quantity of radiation damage.

4.2 Formulation of the model

We consider the model in which the temperature dependence can be interpreted as a result the of phase transition between two plastic deformation structure levels that is characterized by specific values of the athermal stress component.

Changing yield strength σ in dependence on temperature T is characterized by derivative $d\sigma/dT$. As the phase transition is considered between two plastic deformation structure levels then this function must take the form that can be approximated by a parabolic dependence on σ. Such dependence has to be equal zero when yield strength coincides with theoretical quantities of the first (high temperature) $\sigma_{1\mu}^{th}$ and the second (low temperature) $\sigma_{2\mu}^{th}$ athermal plateau. As a result, it can be written a phenomenological equation of the structure phase transition in question as follows

$$\frac{d\sigma}{dT} = \frac{(\sigma_{1\mu}^{th} - \sigma)(\sigma_{2\mu}^{th} - \sigma)}{(\sigma_{1\mu}^{th} - \sigma_{2\mu}^{th})\Delta T_0}, \tag{4.1}$$

where ΔT_0 - the specific temperature interval in that yield strength increases occur on the magnitude of the thermo activated component of flow stress.

Solution of Eq. (4.1) takes the form:

$$\sigma(T) = \frac{\sigma_{1\mu}^{th} + \sigma_{2\mu}^{th}}{2} - \frac{\sigma_{1\mu}^{th} - \sigma_{2\mu}^{th}}{2} \tanh\left(\frac{T - T_c}{2\Delta T_0}\right), \tag{4.2}$$

where T_c is temperature corresponding to the average value of athermal stresses of the high temperature and low temperature plateau.

To describe the yield strength experiment dependences of the irradiated materials on temperature it is convenient to rewrite Eq. (4.2) as follows

$$\sigma(T) = \sigma_c - \sigma_m \tanh\left(\frac{T - T_c}{2\Delta T_0}\right), \tag{4.3}$$

where $\sigma_c = (\sigma_{1\mu}^{th} + \sigma_{2\mu}^{th})/2$, $\sigma_m = (\sigma_{1\mu}^{th} - \sigma_{2\mu}^{th})/2$.

Empirical parameters σ_c, σ_m, ΔT_0, T_c of the model have next phenomenological meaning. Temperature ΔT_0 is connected with activation energy Q_e of the plastic deformation transition on a higher structure level after that the material goes to a stage of radiation embrittlement: $Q_e = \Delta T_0 k_B/2$ where k_B is Boltzmann constant. Parameters σ_c and T_c are stress and temperature of the transition, respectively, between the structure levels of plastic deformation of irradiated materials, characterized by the known experimental values of athremal stress.

From Eq (4.3) follows at $|(T-T_c)/2\Delta T_0| >> 1$ that an equality $\sigma_m = \sigma_{1\mu}^{th} - \sigma_c$ is valid if $T < T_c$ and the equality $\sigma_m = \sigma_c - \sigma_{2\mu}^{th}$ is fulfilled if $T > T_c$ where $\sigma_{1\mu}^{th}$ and $\sigma_{2\mu}^{th}$ are the theoretical magnitudes of the first (low temperature) and the second (high temperature) athermal plateau, respectively (see Fig.4.1.). This implies that parameter σ_m is connected with the thermo activated component of irradiated materials stress by $\sigma_m = \sigma_{th}^{*}/2$.

4.3 Discussion of model results and experimental data

The values of empirical parameters σ_c, σ_m, ΔT_0, T_c are fit by the best coinciding the values of the function (4.3) for corresponding experience temperatures with experimental values of the material yield strength. Criterion of fitting the empirical parameter values has been minimization of a quadratic deviation sum of the yield strength experiment values from ones calculated by Eq (4.3) at corresponding experience temperatures for all of the specific materials.

0X18H10T steel samples have been irradiated by (e,γ) - beams with energy of 225 MeV up to dose of 10^{25} el/cm^2 at temperatures of 170-190°C. For mechanical experiences, the planar samples of test portion sizes of 10×2×0.3 mm have been experienced in vacuum at temperatures of 20-1200°C with deformation velocity of 0.003 c^{-1}[15]. The copper samples of vacuum - induced remelting (purity of 99.98) have been irradiated by γ-beams with energy of 225 MeV up to fluence of 0.1 dpa [15]. The nickel samples have been irradiated by electrons with energy of 225 MeV up to dose of 10^{19} el/cm^2 [11]. The 15X2MΦA steel samples have been irradiated by neutrons up to fluence of $3{\cdot}10^{20}$ neutron/cm^2 [11, 12]. The vanadium samples (purity of 99.9) have been irradiated by high energy (of 225 MeV) (e,γ) - beams up to fluence of 0.01 dpa [12, 15]. Chromium single crystals have been irradiated by (e,γ) - beams up to fluence of 10^{25} el/cm^2 [15]

The summary experimental results for the yield strength temperature dependence of the irradiated materials are shown with Fig.4.2. (fcc lattice materials) and Fig.4.3. (bcc lattice materials) where experimental points are marked by corresponding labels.

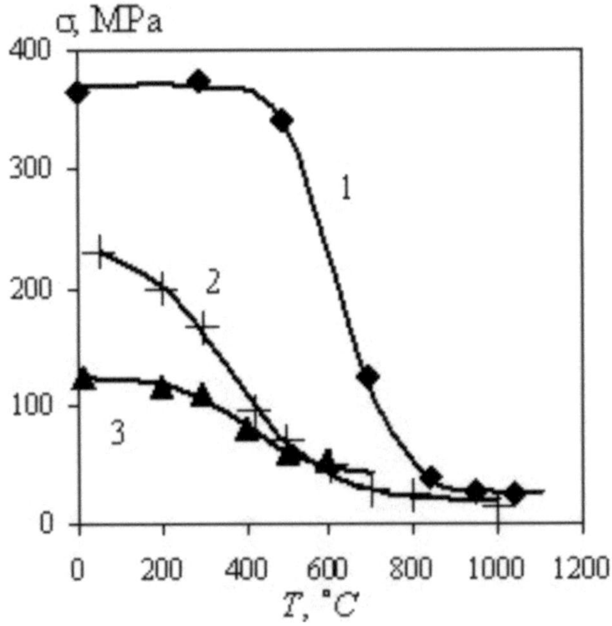

Fig. 4.2. Yield strength temperature dependences of irradiated fcc-materials (point label – experiment, solid lines - theoretical plots calculated by Eq (4.3)). 1 – X18H10T steel; 2 – nickel; 3 – copper.

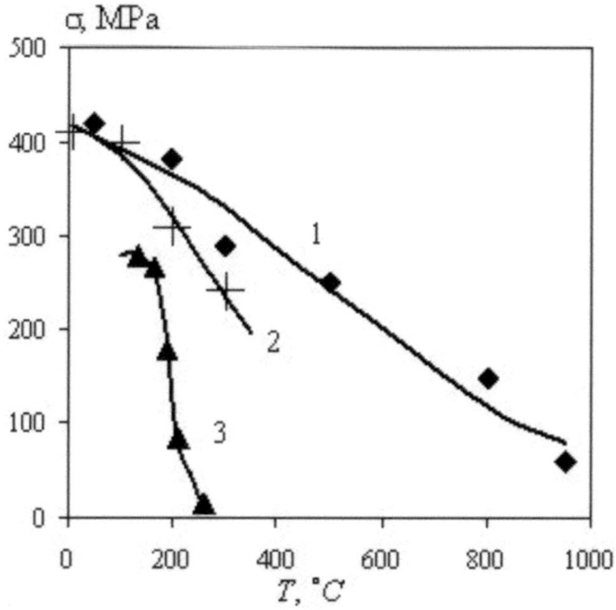

Fig. 4.3. Yield strength temperature dependences of irradiated bcc-materials: 1 - vanadium; 2 - chromium; 3 - 15X2MΦA steel.

There are the approximating function (4.3) values of empirical parameters for the different materials, relative error ε of approximation and confidence quantity R^2 of approximation (determination coefficient) for all of the dependences in the table. The determination coefficient is close to unit. It means good agreement the proposed theoretical dependences with experimental data for all of the considered materials in the wide experience temperature interval. It should be noted that the relative errors for experimental data to be approximated by the function (4.3) in the case of main fcc-materials (X18H10T steel, copper) are lower than in the case of main bcc-materials (vanadium, chromium).

Material	σ_c, MPa	σ_m, MPa	ΔT_0, °C	T_c, °C	ε, %	R^2
0X18H10T steel	197.71	171.99	61.7284	634.87	1.457	0.9997
copper	87.76	35.48	58.1395	376.70	1.799	0.9965
nickel	132.35	113.78	116.2791	357.74	4.100	0.9972
15X2MΦA steel	148.78	135.35	12.9534	196.45	1.083	0.9996
vanadium	241	0.9295	243.9024	505	9.158	0.9638
chromium	277	0.5812	91.4077	251	2.848	0.9810

Table 1. Empirical parameters of the dependence (4.3).

Also the yield strength temperature dependences of no irradiated materials can be approximated by Eq. (4.3) reasonably enough. For instance, the empirical parameters of no irradiated X18H10T steel are $\sigma_c = 59.45$MPa, $\sigma_m = 26.1$ MPa, $T_c = 771.65$ °C, $\varepsilon = 4.405$ %, $R^2 = 0.9814$. Further, it is given the results of comparison of increments for thermo activated $\Delta\sigma^*$, and athermal high temperature $\Delta\sigma_{1\mu}$ and low temperature $\Delta\sigma_{2\mu}$ of stress components obtained experimentally (A) and by a theoretical calculation (B) under radiation up to dose of 10^{25} el/cm^2 for X18H10T steel.

		A	B
$\Delta\sigma_{1\mu}$ MPa	…………………	285	284.149
$\Delta\sigma_{2\mu}$ MPa	…………………	–10	–7.631
$\Delta\sigma^*$ MPa	…………………	295	291.78

According to the data shown at Fig. 4.4., the essential yield strength increment of X18H10T steel is observed after radiarion.

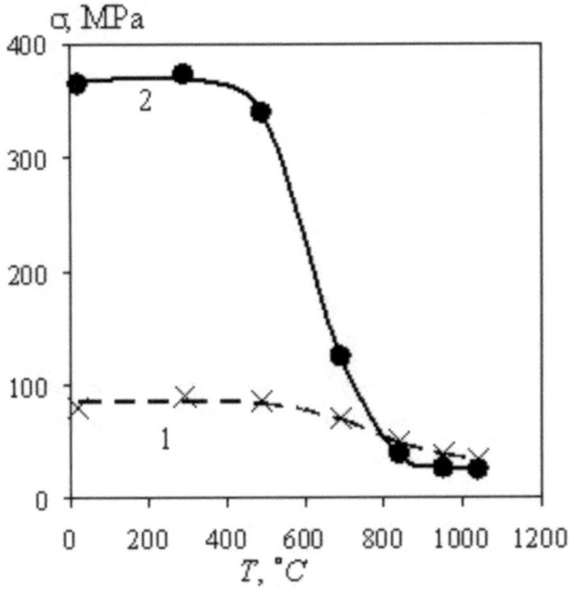

Fig. 4.4. Yield strength temperature dependence of austenic X18H10T steel: 1 – no irradiated, 2 – irradiated by (e,γ) – beams up to dose of 10^{25} el/cm^2. Point labels are experimental; the lines are plots of theoretical dependence calculated by Eq. (4.3).

5. Conclusion

In this chapter, three models are proposed to describe the properties of irradiated deformed polycrystalline materials.

The first model (see section 2) describes the dose dependence of the yield strength of the irradiated material. It is generalization of the model known earlier (see for instance [5]), as the recombination effects of the radiation – induced vacancy and interstitial barriers and their clusterization are taken into account in it. Within the framework of the model formulated, it is found that the saturation quantity of the yield strength decreases with increasing both the intensity of the mutual barrier recombination and the clusterization intensity. It follows to note that in spite of model's assumption for the mean sizes of the radiation-induced barriers d_i and the shear modulus μ to be independent the shear modulus increases practically owing to the radiation point defects come on the dislocation and reduce the mean segment length of the dislocation and enhance a degree of the dislocation anchorage. Because of this it can be expected for the obtained dose dependence of the yield strength of the irradiated material to be modified.

The second model describes barrier hardening polycrystalline materials. It is constructed with taking into account interaction of vacancy and interstitial barrier types. In the frame of the proposed model, it can be estimated both contributions to yield strength increment from different type barriers and its total value in dependence on dose.

The third model gives possibility to describe the yield strength dependence of the irradiated materials on experience temperature on a quantitative level. It is based on mechanism of yield strength change as the phase transition between two plastic deformation structure levels characterized by certain values of the athermal stress component. The calculations show radiation promotes the transition of plastic deformation on the higher structure level after that the material undergoes radiation embrittlement. General features found permit to forecast embrittlement temperature intervals of reactor materials in dependence on their mechanical properties.

6. References

[1] Tompson M W Defects and Radiation Damage in Metals (London:Cambridge Univ. Press, 1969)

[2] V. Naundorf, "Diffusion in metals and alloys under irradiation," International Journal of Modern Physics B, 6 No. 18 2925--2986 (1992).

[3] A. M. Parshin, I. M. Neklyudov, N.V. Kamyshanchenko, A. N. Tikhonov, et al., @Physics of radiation phenomena and radiation material@ (Publishing BSU, Belgorod, 1998) [in Russian].

[4] N. M. Ghoniem, J. Alhajji and Garner F.A., "Hardening of irradiated alloys due to the simultaneous formation of vacancy and interstitial loops," in @Effect of radiation on materials@ (ASTM, Philadelphia, 1982), pp.1054--1087.

[5] K. Khavanchik, D. Senesh and V. A. Shchegolev, "On saturation of yield strength of copper iiradiated by energy charged particles," PhHOM, No. 2, 5--10 (1989) [in Russian].

[6] ITER Interim Structural Design Criteria (SDC-IC) (ITER Doc.IDoMS G 74 MA 8 01-05-28 W02, 2001).

[7] G.M. Kalinin, B.S. Rodchenkov and V.A. Pechenkin, "Specification of stress limits for irradiated 316L(N)-IG steel in ITER structural design criteria," Journal of Nuclear Materials 329–333, 1615--1618 (2004).

[8] V. C. Neustroyev, Z. E. Ostrovsky, E. V. Boyev and S. V. Belozerov, "Influence of helium accumulation in austenitic steel on evolution of microstructure and radiation damageability of WWER reactor vessel internals materials," in Abstracts of VII Russian conference on reactor material science (NIIAR, Dmitrovgrad, 2007), p 117 [in Russian].

[9] V.V. Krasil'nikov and S.E. Savotchenko, Russian Metallurgy (Metally), Vol. 2009,No. 2, pp. 172-178.

[10] Svetukhin V.V., Sidorenko O.G., Golovanov V.N. and Suslov D.N., Fiz. Khim. Obrab. Mater., 2005, No. 3, pp. 15-20.

[11] A.M. Parshin, I.M. Neklyudov, N.V. Kamyshanchenko, et al., Physics of Radiation Phenomena and Radiation Material Science (Publishing BSU, Belgorod, 1998) [In Russian].

[12] A. A. Parkhomenko, Microstructure and Radiation Embrittlement of Nickel and OX16P15M3Б Steel, Electron microscopy and Durability of Crystals, No. 9, 103 (1998) [In Russian].

[13] M. P. Zeidlits, L. S. Ozhigov, A. A. Parkhomenko, et al. @Influence High Energy Electron Radiation on Radiation Hardening of Nickel, Vanadium, Iron and Their Alloys, Problems of Atomic Science and Technology. Series: Phys. Rad. Dam., and Rad. Mater. Sc., No. 1 (2), 36 (1975) [In Russian].

[14] V. F. Zelensky, I. M. Neklyudov, L. S. Ozhigov, et al., Utilization of Electron Accelerators for Simulation and Studies of Radiation Effects on Mechanical Properties of Fusion Reactor Materials, J. Nucl. Mater., 207, 280 (1993).

[15] I. M. Neklyudov, V. N. Voevodin, L. S. Ozhigov, et al., Temperature Dependences of Mechanical Properties and Radiation Hardening of Materials, Izv. TulSU. Ser. Physics, No. 4, 87 (2004) [In Russian]

[16] I.M. Neklyudov, I.S. Ozhigov, A.A. Parkhomenko. Utilization of the charge particles of accelerators for simulation of reactor damage effects V.F. Zelensky, // J. Nucl. Mater. - 1993. - Vol. 207. - pp.280-285.

[17] V.F. Zelensky, I.S .Ozhigov, I.M. Neklyudov, Proc. Int. Conf. Irradiation behavior of metallic for fast reactor core components. - France, Corce. 1979. - pp.131-160.